SpringerBriefs in Applied Sciences and Technology

Wen Yu · Suresh Thenozhi

Active Structural Control with Stable Fuzzy PID Techniques

 Springer

Wen Yu
Departamento de Control Automatico
CINVESTAV-IPN
Mexico City
Distrito Federal
Mexico

Suresh Thenozhi
Departamento de Control Automatico
CINVESTAV-IPN
Mexico City
Distrito Federal
Mexico

ISSN 2191-530X ISSN 2191-5318 (electronic)
SpringerBriefs in Applied Sciences and Technology
ISBN 978-3-319-28024-0 ISBN 978-3-319-28025-7 (eBook)
DOI 10.1007/978-3-319-28025-7

Library of Congress Control Number: 2015958543

Printed on acid-free paper

This Springer imprint is published by SpringerNature
The registered company is Springer International Publishing AG Switzerland

Contents

Chapter 1
Introduction

Abstract This chapter provides an overview of active structure control. Some typical applications of structural control systems with respect to earthquakes are discussed.

Keywords Structural control · Vibration attenuation · Earthquake

In the past few decades, there have been several applications of vibration control systems in building structures. The implementation of control devices in buildings and bridges has been investigated by many researchers worldwide. It is reported that over 16, 000 building structures around the world have been protected using anti-seismic systems [1]. Passive control, such as seismic isolator, is one of the widely implemented technique. Japan has about 18 % of all earthquakes on the planet, of magnitude 7 or more. Over 5, 000 buildings in Japan have already been protected by seismic isolators. A brief review on the application of seismic isolations in Japan can be found in [2, 3]. A recent survey on the performance evaluation of the seismic isolation is presented in [4]. It is reported that the USC University Hospital, world's first base isolated hospital survived the 1994 Northridge earthquake (Mag-6.7 caused $20 billion in damage), without any damage. Some other examples are, the world's largest base-isolated computer center survived the 1995 Kobe earthquake (Mag-7.1 caused $150 billion in damage) and the 7-story Christchurch Women's Hospital in the South Island of New Zealand survived the February 2011 Christchurch earthquake [5].

Active devices, such as active mass dampers (AMDs), are implemented to mitigate human-induced vibrations on the in-service footbridge. With a mass ratio of 0.2 %, the vibration is reduced by 60–80 % [6]. The Shanghai World Financial Center in China, holds two AMDs below its observation floors to reduce the building vibration during windstorms. During earthquakes, complete active control is disabled, hence works as tuned mass dampers (TMDs) [7]. The first full-scale implementation of semi-active magneto rheological (MR) damper is applied to the Miraikan Building in Tokyo [8]. The Osaka Applause Tower in Osaka, Japan is equipped with an AMD. This damper system has attenuated the structural response due to the typhoon on July 26th, 1996 and earthquake on April 22nd, 1998 [8]. Some practical applications of active and semi-active vibration control of buildings are reported in [9, 10]. The

© The Author(s) 2016

W. Yu and S. Thenozhi, *Active Structural Control with Stable Fuzzy PID Techniques*, SpringerBriefs in Applied Sciences and Technology, DOI 10.1007/978-3-319-28025-7_1

Island Tower Sky Club in Fukuoka City, Japan, uses different kinds of dampers for protection against wind and seismic forces. A hybrid base isolation system is applied to reduce the building response during large earthquakes. Its towers are interconnected using dampers to reduce the overturning effect [11]. A state-of-the-art-review of the behavior of isolated bridges to seismic excitation is presented in [12]. More information on buildings with vibration control systems can be found in [13].

On March 11th, 2011, a mega-thrust earthquake of moment magnitude 9.0 occurred offshore NE Japan. This is the most devastating earthquake in Japan after the 1923 Great Kanto earthquake. The building experienced strong shaking due to the long-period ground motions (0.015–0.1 Hz). Despite these conditions, it is reported that many buildings equipped with structural control system has performed well. Many research results about this earthquake are presented in "International symposium on engineering lessons learned from the 2011 Great East Japan Earthquake, UC Berkeley, March 1–4, 2012". In [14], the response of eight buildings with structural control systems are reported and found that the control system responded well. It states that a 54-story steel building experienced a maximum displacement of 0.50 m, which would have reached 0.70 m in the absence of the control. The Ishinomaki Red Cross Hospital in Ishinomaki, situated near to the epicenter, had a good controlled response [5]. Miwada et al. [15] shows that the base isolated systems using high damping rubber bearings in Miyagi and Chiba prefectures performed well during the Tohoku earthquake. Takewaki et al. [16] shows that some super high-rise buildings with viscoelastic dampers like high-hardness rubber dampers controlled the floor vibration effectively.

Beside these successful performances, some buildings with anti-seismic systems are damaged during Tohoku-Oki earthquake. A damage to a base isolated building with a lead damper is indicated in [17]. It is also reported that the structures designed by post-1990 code are few damaged by the ground motion, whereas many structures designed by post-1995 code, using rubber bearings and dampers are severely damaged [16].

Also some control system did not respond to this earthquake. For example, the damper system (three TMDs) in Taipei 101 skyscraper acted during the 2005 Typhoon Long Wang, the 2008 Wenchuan Earthquake, and the 2010 Typhoon Fanapi. However it did not respond during the 2011 Tohoku-Oki earthquake, because it was not excited under those circumstances [18]. From an engineering point of view, the observations of these events will be useful for researchers to identify relevant research questions about the structure safety and for improving structure resilience against these natural hazards and it is also important to investigate the performance of the anti-seismic systems to different excitations.

References

1. A. Martelli, M. Forni, G. Panza, Features, recent application and conditions for the correct use of seismic isolation systems. Seismic Control Syst. Des. Perform. Assess. **120**, 15–27 (2011)
2. T. Fujita, Seismic isolation of civil buildings in Japan. Prog. Struct. Eng. Mater. **1**(3), 295–300 (1998)
3. S. Kawamura, R. Sugisaki, K. Ogura, S. Maezawa, S. Tanaka, Seismic isolation retrofit in japan, in *Proceedings of the 2th World Conference on Earthquake Engineering* (2000), pp. 1–8
4. I.G. Buckle, R.L. Mayes, Seismic isolation: history, application, and performance-a world view. Earthq. spectra. **6**, 161–201 (2012)
5. R.L. Mayes, A.G. Brown, D. Pietra, Using seismic isolation and energy dissipation to create earthquake-resilient buildings. Bull. N. Z. Soc. Earthq. Eng. **45**, 117–122 (2012)
6. C.M. Casado, I.M. Díaz, J.D. Sebastián, A.V. Poncela, A. Lorenzana, Implementation of passive and active vibration control on an in-service footbridge. Struct. Control Health Monit. **20**, 70–87 (2013)
7. X. Lu, P. Li, X. Guo, W. Shi, J. Liu, *Vibration control using ATMD and site measurements on the Shanghai World Financial Center Tower* (Struct. Des. Tall. Spec, Build, 2012)
8. Z.H. Dong, J.Y. Yuan, Vibration control device and its performance under wind load in high-rise buildings. Appl. Mech. Mater. **166–169**, 1358–1361 (2012)
9. F. Casciati, J. Rodellar, U. Yildirim, Active and semi-active control of structures–theory and applications: a review of recent advances. J. Intell. Mater. Syst. Struct. **23**(11), 1181–1195 (2012)
10. Y. Ikeda, Active and semi-active vibration control of buildings in Japan–practical applications and verification. Struct. Control Health Monit. **16**, 703–723 (2009)
11. A. Nishimura, H. Yamamoto, Y. Kimura, H. Kimura, M. Yamamoto, A. Kushibe, Base-isolated super high-rise RC building composed of three connected towers with vibration. Struct. Concr. **12**(2), 94–108 (2011)
12. M.C. Kunde, R.S. Jangid, Seismic behavior of isolated bridges: a-state-of-the-art review. Electron J. Struct. Eng. **3**, 140–170 (2003)
13. T.K. Datta, A state-of-the-art review on active control of structures. ISET J. Earthq. Technol. **40**, 1–17 (2003)
14. K. Kasai, W. Pu, A. Wada, Responses of Controlled Tall Buildings in Tokyo Subjected to the Great East Japan Earthquake, in *Proceedings of the International Symposium on Engineering Lessons Learned from the 2011 Great East Japan Earthquake* (2012), pp. 1099–1109
15. G. Miwada, O. Yoshida, R. Ishikawa, M. Nakamura, Observation Records of Base-Isolated Buildings in Strong Motion Area During the 2011 Off the Pacific Coast of Tohoku Earthquake, *Proceedings of the International Symposium on Engineering Lessons Learned from the 2011 Great East Japan Earthquake*, 2012, pp. 1017–1024
16. I. Takewaki, S. Murakami, K. Fujita, S. Yoshitomi, M. Tsuji, The 2011 off the Pacific coast of Tohoku earthquake and response of high-rise buildings under long-period ground motions. Soil Dyn. Earthq. Eng. **31**(11), 1511–1528 (2011)
17. M. Motosaka, K. Mitsuji, Building damage during the 2011 off the Pacific coast of Tohoku Earthquake. Soils Found. **52**(5), 929–944 (2012)
18. K.C. Chen, J.H. Wang, B.S. Huang, C.C. Liu, W.G. Huang, Vibrations of the TAIPEI 101 skyscraper caused by the 2011 Tohoku earthquake. Japan. Earth Planets Space **64**, 1277–1286 (2013)

Chapter 2
Active Structural Control

Abstract This chapter provides an overview of building structure modeling and control. It focuses on different types of control devices, control strategies, and sensors used in structural control systems. It also discusses system identification techniques and some important implementation issues.

Keywords Structural control · Vibration attenuation · Mathematical modeling

2.1 Introduction

Structural vibration can be generally controlled in two ways: (1) by constructing the buildings using smart materials [2]; (2) by adding controlling devices like dampers, isolators, and actuators to the building [3–5]. In this work, we only discuss the latter case, where the structural dynamics are modified favorably by adding active devices. The performance of a structural control system depends on various factors including excitation type (e.g., earthquakes and winds), structural characteristics (e.g., degree of freedom, natural frequency, and structure nonlinearity), control system design (e.g., type and number of devices, placement of devices, system model, and the control algorithm), etc. [6]. In active control, the structural response under the input excitations are measured using sensors and an appropriate control force, calculated by a preassigned controller is used to drive the actuators for suppressing the unwanted structure vibrations.

Due to the popularity and importance of structural control, a number of textbooks [7, 8] and review papers have been presented. A brief review was presented by Housner et al. [2] in 1997, which discusses the passive, active, semi-active, and hybrid control systems and explores the potential of control theory in structural vibration control. It explains different types of control devices and sensors used in structural control. The paper concludes with some recommendations for future research.

W. Yu and S. Thenozhi, *Active Structural Control with Stable Fuzzy PID Techniques*, SpringerBriefs in Applied Sciences and Technology, DOI 10.1007/978-3-319-28025-7_2

A recent survey on active, semi-active, and hybrid control devices and some control strategies for smart structures were presented in [9, 10]. Some reviews were carried out with particular emphasis on active control [11–15], on semi-active control [16], and on control devices [17–19]. This shows that a significant progress has been made on most aspects of the structural control in the past few decades.

While there is no doubt about the advance, there still exist some areas which need more exploration. During the seismic excitation the reference where the displacement and velocity sensors are attached will also move, as a result the absolute value of the above parameters cannot be sensed. Alternatively, accelerometers can provide inexpensive and reliable measurement of the acceleration at strategic points on the structure. Most of the controllers use the displacement and velocity as its input variable, which are not easy to obtain from the acceleration signal with simple integration. Application of the state observers is impossible if the system parameters are unknown. Similarly, parameter uncertainty may be a problem for some control designs. There are different techniques available for identifying building parameters [20]. But these parameters may change under different load conditions. However, these control laws would be more applicable to real buildings if they could be made adaptive and robust to system uncertainty.

The active devices have the ability to add force onto the building structure. If the controller generates unstable dynamics, it can cause damage to the building. So it is important to study the stability of the controller. Only a few structural controllers such as H_∞ and sliding mode controller consider the stability in their design, whereas the other control strategies do not. Also, there is a lack of experimental verification of these controllers. Some other areas that demands attention are the time-delay present in the actuator mechanism, actuator saturation, and the optimal placement of sensors and actuators. The implementation of a controller will be challenging if these issues were not resolved. The motivation for the work presented in this book is to push forward the performance and capabilities of the structural vibration control system by acknowledging the above issues.

The objective of structural control system is to reduce the vibration and to enhance the lateral integrity of the building due to earthquakes or large winds, through an external control force [21]. In active control system, it is essential to design one controller in order to send an appropriate control signal to the control devices so that it can reduce the structural vibration. The control strategy should be simple, robust, fault tolerant, need not be an optimal, and of course must be realizable [22].

This chapter provides an overview of building structure modeling and control. It focuses on different types of control devices and control strategies used in structural control systems. This chapter also discusses system identification techniques and some important implementation issues, like the time-delay in the system, state estimation, and optimal placement of the sensors and control devices. A detailed version of this chapter can be found in [23].

2.2 Modeling of Building Structures

2.2.1 Models of Building Structures

In order to derive a dynamic model of a building structure, it is important to know the behavior and impact of the excitations on the buildings, such as strong wind and seismic forces. The force exerted by the earthquake and wind excitation on the structure is shown in Fig. 2.1. An earthquake is the result of a sudden release of energy in the Earth crust that creates seismic waves. The building structure oscillates with the ground motion caused by these seismic waves and as a result the structure floor masses experience the inertia force. This force can be represented as

$$f = -m\ddot{x}_g \tag{2.1}$$

where m is the mass and $\ddot{x}_g$ is the ground acceleration caused by the earthquake.

The movement of the structure depends on several factors like the amplitude and other features of the ground motion, the dynamic properties of the structure, the characteristics of the materials of the structure and its foundation (soil-structure interaction). A civil structure will have multiple natural frequencies, which are equal to its number of degree-of-freedom (DOF). If the frequency of the motion of the ground is close to the natural frequency of the building, resonance occurs. As a result, the floors may move rigorously in different directions causing inter-story drift, the relative translational displacement between two consecutive floors. If the building drift value or deformation exceeds its critical point, the building damages severely. Small buildings are more affected by high-frequency waves, whereas the large structures or high-rise buildings are more affected by low-frequency waves. The major part of the structure elastic energy is stored in its low order natural frequencies, so it is important to control the structure from vibrating at those frequencies [24].

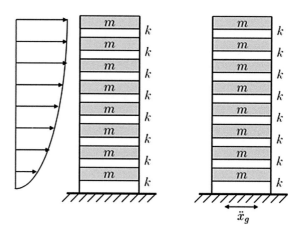

Fig. 2.1 a Wind excitation; **b** earthquake excitation

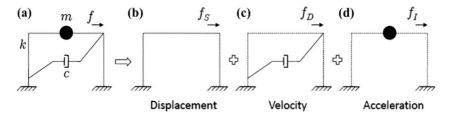

Fig. 2.2 a Structure; **b** stiffness component; **c** damping component; **d**mass component

In the case of high-rise flexible buildings, strong winds cause sickness or psychological responses like anxiety to the occupants and also may damage the fragile items. When the vibrations of taller buildings due to the high wind exceed a limit of $0.15\,\mathrm{ms}^2$, humans may feel uncomfortable [18]. As a result, the main objective of structural control is to reduce the acceleration response of buildings to a comfortable level. The force exerted by the wind on a building structure can be represented as [25];

$$F_w(h_i, t) = \Upsilon(h_i)v(t) \tag{2.2}$$

where $v(t)$ is the dynamic wind speed and $\Upsilon(h_i)$ has the following expression.

$$\Upsilon(h_i) = \rho_a \mu_p \mu_h \Delta_w(h_i)v_m \tag{2.3}$$

where ρ_a is the air density, μ_p is the wind pressure coefficient, $\Delta_w(h_i)$ is the windward area of the structure at elevation h_i, and v_m is the mean wind speed. The wind profile coefficient μ_h can be expressed as

$$\mu_h = (0.1h_i)^{2\alpha_a} \tag{2.4}$$

where α_a is a positive constant.

It is worth to note that the main difference between the effects of earthquake and wind forces on a structure is that, the earthquake causes internally generated inertial force due to the building mass vibration, whereas wind acts in the form of externally applied pressure.

A single-degree-of-freedom (SDOF) structure can be modeled using three components: the mass component m, the damping component c, and the stiffness component k [26], which is shown in Fig. 2.2. The stiffness component k can be modeled as either a linear or a nonlinear component, in other words elastic or inelastic, respectively [13]. Usually the mass is considered as a constant. When an external force f is applied to a structure, it produces changes in its displacement $x(t)$, velocity $\dot{x}(t)$, and acceleration $\ddot{x}(t)$.

Consider a simple building structure, which can be modeled by [26],

$$m\ddot{x} + c\dot{x} + kx = f_e \tag{2.5}$$

Fig. 2.3 Mechanical model of a n-DOF building structure

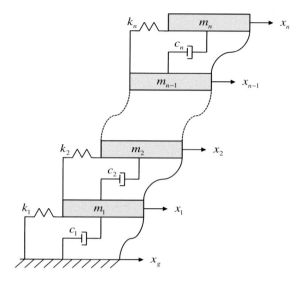

where m is the mass, c is the damping coefficient, k is the stiffness, f_e is an external force applied to the structure, and x, $\dot{x}$, and $\ddot{x}$ are the displacement, velocity, and acceleration, respectively.

A model for a linear multistory structure with n-degree-of-freedom (n-DOF) is shown in Fig. 2.3. Here, it is assumed that the mass of the structure is concentrated at each floor. Neglecting gravity force and assuming that a horizontal force is acting on the structure base, the equation of motion of the n-floor structure can be expressed as [13],

$$M\ddot{\mathbf{x}} + C\dot{\mathbf{x}} + \mathbf{f}_s = -\mathbf{f}_e \tag{2.6}$$

For unidirectional motion, the parameters can be simplified as [13]:

$$M = \begin{bmatrix} m_1 & 0 & \cdots & 0 \\ 0 & m_2 & \cdots & \vdots \\ \vdots & \vdots & \ddots & \vdots \\ 0 & 0 & \cdots & m_n \end{bmatrix} \in \mathfrak{R}^{n \times n}, \quad C = \begin{bmatrix} c_1 + c_2 & -c_2 & \cdots & 0 & 0 \\ -c_2 & c_2 + c_3 & \cdots & \vdots & \vdots \\ \vdots & \vdots & \ddots & \vdots & \vdots \\ \vdots & \vdots & & c_{n-1} + c_n & -c_n \\ 0 & 0 & \cdots & -c_n & c_n \end{bmatrix} \in \mathfrak{R}^{n \times n},$$

$$\tag{2.7}$$

$\mathbf{x} \in \mathfrak{R}^n$, $\mathbf{f}_s = \begin{bmatrix} f_{s,1} \cdots f_{s,n} \end{bmatrix} \in \mathfrak{R}^n$ is the structure stiffness force vector, and $\mathbf{f}_e \in \mathfrak{R}^n$ is the external force vector applied to the structure, such as earthquake and wind excitations.

If the relationship between the lateral force $\mathbf{f}_s$ and the resulting deformation $\mathbf{x}$ is linear, then $\mathbf{f}_s$ is

$$
\mathbf{f}_s = K\mathbf{x}, \text{ where } K = \begin{bmatrix} k_1 + k_2 & -k_2 & \cdots & 0 & 0 \\ -k_2 & k_2 + k_3 & \cdots & \vdots & \vdots \\ \vdots & \vdots & \ddots & \vdots & \vdots \\ \vdots & \vdots & \cdots & k_{n-1} + k_n & -k_n \\ 0 & 0 & \cdots & -k_n & k_n \end{bmatrix} \in \mathfrak{R}^{n \times n} \qquad (2.8)
$$

If the relationship between the lateral force $\mathbf{f}_s$ and the resulting deformation $\mathbf{x}$ is nonlinear, then the stiffness component is said to be inelastic [26]. This happens when the structure is excited by a very strong force, that deforms the structure beyond its limit of linear elastic behavior. Bouc–Wen model gives a realistic representation of the structural behavior under strong earthquake excitations. The force-displacement relationship of each of the stiffness elements (ignoring any coupling effects) agrees the following relationship [27]:

$$
f_{s,i} = \epsilon k_i x_i + (1 - \epsilon) k_i \eta \varphi_i, \quad i = 1 \cdots n \qquad (2.9)
$$

where the first part is the elastic stiffness and the second part is the inelastic stiffness, k_i is the linear stiffness defined in (2.8), ϵ and η are positive numbers, and φ_i is the nonlinear restoring force which satisfies

$$
\dot{\varphi}_i = \eta^{-1} \left[\delta \dot{x}_i - \beta |\dot{x}_i| |\varphi_i|^{p-1} \varphi_i + \gamma \dot{x}_i |\varphi_i|^p \right] \qquad (2.10)
$$

where δ, β, γ, and p are positive numbers. The Bouc–Wen model has hysteresis property. Its input displacement and the output force is shown in Fig. 2.4. The dynamic properties of the Bouc–Wen model has been analyzed in [28].

In the case of closed-loop control systems, its input and output variables may respond to a few nonlinearities. From the control point of view, it is crucial to investigate the effects of the nonlinearities on the structural dynamics.

The Bouc–Wen model represented in (2.9) and (2.10) is said to be bounded input-bounded-output (BIBO) stable, if and only if the set Ω_{bw} with initial conditions $\varphi(0)$ is nonempty. The set Ω_{bw} is defined as: $\varphi(0) \in \mathfrak{R}$ such that f_s is bounded for all

Fig. 2.4 Hysteresis loop of Bouc–Wen model

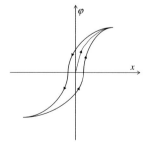

Table 2.1 Stability of Bouc–Wen model with different δ, β, γ.

Case	Conditions	Ω_{bw}	Upper bound of $|\varphi(t)|$
1	$\delta > 0$, $\beta + \gamma > 0$ and $\beta - \gamma \geq 0$	$\mathfrak{R}$	$\max\left(|\varphi(0)|, \varphi_a\right)$
2	$\delta > 0$, $\beta - \gamma < 0$ and $\beta \geq 0$	$[-\varphi_b, \varphi_b]$	$\max\left(|\varphi(0)|, \varphi_a\right)$
3	$\delta < 0$, $\beta - \gamma > 0$ and $\beta + \gamma \geq 0$	$\mathfrak{R}$	$\max\left(|\varphi(0)|, \varphi_b\right)$
4	$\delta < 0$, $\beta + \gamma < 0$ and $\beta \geq 0$	$[-\varphi_a, \varphi_a]$	$\max\left(|\varphi(0)|, \varphi_b\right)$
5	$\delta = 0$, $\beta + \gamma > 0$ and $\beta - \gamma \geq 0$	$\mathfrak{R}$	$|\varphi(0)|$
6	All other conditions	$\emptyset$	Unbounded

C^1 input signal, and x with fixed values of parameters δ, β, γ, and p, φ_a and φ_b are defined as

$$\varphi_a = \sqrt[p]{\frac{\delta}{\beta + \gamma}}, \quad \varphi_b = \sqrt[p]{\frac{\delta}{\gamma - \beta}} \tag{2.11}$$

For any bounded input signal x, the corresponding hysteresis output f_s is also bounded. On the other hand if $\varphi(0) \in \Omega_{bw} = \emptyset$, then the model output f_s is unbounded. Table 2.1 shows how the parameter δ, β, γ, affect the stability property of the Bouc–Wen model.

In the case of n-DOF structures, the nonlinear model can be modified as

$$M\ddot{x}(t) + C\dot{x}(t) + F_s(x(t), \dot{x}(t)) = -M\Lambda\ddot{x}_g(t) \tag{2.12}$$

where $\Lambda \in \mathbb{R}^{n \times 1}$ denotes the influence of the excitation force.

2.2.2 Estimation and Sensing of Structure Parameters

Sensor and actuator placement. The optimal placement is concerned with placement of the sensing and controlling devices in preselected regions in order to closely perform the measurement and control operation of the structure vibration optimally. The actuator and sensor play an important role in deciding the system's controllability and observability, respectively. So it is important to perform an optimal placement of the sensors and actuators such that the controllability and observability properties of all or selected modes are maximized. Due to the above-mentioned reasons and importance, a number of studies are carried out about the optimal placements of devices [29, 30]. A survey on the optimal placement of control devices can be found in [10].

In [31], the actuator and sensor location performance index is calculated between the γ_{WZ}^2 and γ_{UY}^2 Hankel singular values. A nonnegative correlation coefficient κ is defined as

$$\kappa^2 = \frac{(\gamma_{WZ}^2)^T \gamma_{UY}^2}{\|\gamma_{WZ}^2\|_2 \|\gamma_{UY}^2\|_2} \tag{2.13}$$

where γ_{WZ}^2 and γ_{UY}^2 represents the Hankel singular values of the transfer functions G_{WZ} and G_{UY}, respectively. Here U and W are the inputs to the system and Y and Z are the outputs of the system. As per the above equation, the maximal performance is obtained with a better controllability and observability properties when κ reaches a maximal value; $\kappa = 1$, which is achieved when $\gamma_{UY}^2 = \gamma_{WZ}^2$.

A closed-loop optimal location selection method for actuators and sensors in flexible structures is developed by Guney et al. [30], which uses a simple H_∞ controller where the location optimization is performed using a gradient-based unconstrained minimization. Another related work is done in [32] using a H_2 norm-based computation for a reduced model of flexible structures, which considers only the dominant modes. They also proposed one GA for the nonlinear optimization problem for the reduced-order model. A GA is proposed in [33] through the formulation of a discrete and nonlinear optimization problem. Finally, the proposed algorithm is simulated for a 16-story building under 18 different earthquake excitations. In the work [25], it is concluded that the optimal position of actuators depends on the control algorithm, so that different control algorithms or different controllers yield different positions of the actuators.

Sensing. In order to identify the parameters of the civil structures, the dynamic response is studied from its input and output data, and the parameters are estimated using some sort of identification techniques. The inputs are the excitation forces like the earthquake and wind loads, and the outputs are the displacements, velocities, and accelerations corresponding to the input excitation. In practice, it is very difficult to derive an exact system model, so the original problem is to obtain parameters, such that the estimated model responses closely match the output of the building dynamic behaviors. There exists different methods for identification of both linear and nonlinear systems [34].

For the purpose of system identification, the structural system can be represented in many ways, such as ordinary differential equation (ODE), transfer functions, state-space models, and Auto Regressive Moving Average with exogenous input (ARMAX) models [35]. Consider a state-space variable $z = \left[x^T, \dot{x}^T\right]^T \in \Re^{2n}$, then the system described in (2.41) can be represented in state-space form as

$$\dot{z}(t) = Az(t) + Bu(t) + E\ddot{x}_g(t) \tag{2.14}$$

$$y(t) = Hz(t) + Du(t) \tag{2.15}$$

where $A \in \Re^{2n \times 2n}$, $B \in \Re^{2n \times n}$ and $E \in \Re^{2n}$.

$$A = \begin{bmatrix} 0 & I_n \\ -M^{-1}K & -M^{-1}C \end{bmatrix}$$

$$B = \begin{bmatrix} 0 \\ M^{-1}\Upsilon \end{bmatrix}, \quad E = \begin{bmatrix} 0 \\ -\theta \end{bmatrix}$$

Here, the matrices H and D and their dimensions change according to the design demands.

System identification can be broadly classified into parametric and nonparametric identification. In parametric identification, the system parameters like the mass, stiffness, and damping are estimated [36]. Most commonly used algorithms are least squares method, maximum likelihood method, extended Kalman filter, and variations of them [35]. Nonparametric identification determines a system model from the measured data, which is a mathematical function that can approximate the input-output representations sufficiently well [37]. This method is suitable for the systems with infinite number of parameters. Artificial neural network (ANN) is one of the popular nonparametric identification method [38]. Some other known methods are wavelet networks, splines, and neuro-fuzzy models [20].

Identification can also be classified into time-domain and frequency-domain, where the identification takes the form of time series and frequency response functions or spectra, respectively [20, 35]. System identification can be performed either using online or offline techniques. In offline identification, all the data including the initial states must be available before starting the identification process. For example, in the case of building parameter identification, the excitation and the corresponding structure response are recorded and later used for identification. Whereas, the online identification is done immediately after each input-output data is measured. In other words, the online identification is performed parallel to the experiment that is during the structural motion due to seismic or wind loads.

System identification of a linear MDOF structure under ambient excitation using the eigen space algorithm is presented in [39]. The algorithm identifies the damping and stiffness with known mass. In [40], two backpropagation neural networks (BPNN) are used to estimate the stiffness and damping of a 5-story building, where the first one is called emulator NN and the second one is known as the parametric evaluation NN. A modified GA strategy [41] and GA with gradient search [42] is proposed to improve the accuracy and computational time for parameter identification of MDOF structural systems. Sometime, the parameters are identified in the structure equipped with the actuator [6]. On the other hand, identification is performed only for the control devices. In [43], a memory-based learning called lazy recursive learning method based on NN is used to identify the MR damper behavior. The input current to the MR damper is varied and the corresponding damper behavior is modeled.

System identification is sometimes used for modal analysis, where the modal parameters like natural frequencies (ω_n) for different modes, modal shapes, and damping ratios (ζ) of the structures are estimated [20]. One such a simple technique is the analysis using Fourier transform techniques to estimate power spectra from which the modal parameters are estimated [35]. When the input excitation frequency equals the structure natural frequency, the magnitude of the vibration becomes higher. So it is important to estimate these low order natural frequencies and to control the structure from vibrating at those frequencies. A modified random decrement method along with Ibrahim time-domain technique is used for estimating the modal parameters, which uses the floor acceleration [25]. The modal parameters can also be identified using Kalman filter [44].

Parametric identification of a linear structure excited with two orthogonal horizontal components using least-squares identification algorithm is presented in [45]. Here, each floor is considered to have 3-DOF, two displacements (along the x and y axis) and one torsion (rotation around the z axis). In [38], the dynamic state-space model of an earthquake-excited structure is identified using the measured input-output data that is used later for estimating the modal parameters. The system and modal parameters of a linear MDOF structure is estimated in [46]. Here, the equation of motion of the structure is first written in state-space equation of the observable canonical form and then is converted into an ARMAX model for dealing with the noise present in the measured data.

Some works [47] consider the damping matrix C as a Rayleigh damping coefficient matrix, which is found using the modal parameters as given below [26],

$$C = \alpha_R M + \beta_R K \tag{2.16}$$

where the Rayleigh parameters α_R and β_R are calculated using the first and third eigen-frequencies (ω_1 and ω_3), given by

$$\alpha_R = \frac{2\zeta\omega_1\omega_3}{\omega_1 + \omega_3} \text{ and } \beta_R = \frac{2\zeta}{\omega_1 + \omega_3} \tag{2.17}$$

whereas [24] uses the first two lower-order mode frequencies.

In [42], the stiffness of the structure column is estimated using the equation given below

$$k = \frac{12E_y I_m}{L_c^3} \tag{2.18}$$

where E_y is the Young's modulus of elasticity, I_m is the moment of inertia, and L_c is the unsupported length of the column.

A brief review about the identification of nonlinear dynamic structures is presented by Kerschen et al. [20] in 2006. The fundamentals and methods of identification for linear and nonlinear structural dynamic systems are reviewed in [35]. A general survey on system identification is presented in [48] and a review on stochastic identification methods for modal analysis is presented in [41].

Estimation of System States. In order to control the structural dynamics, it is necessary to measure the system states directly using a sensor or indirectly by using a state observer. Some structural control applications use Kalman filter as the observer for estimating the velocity and displacement [49]. A Kalman filter estimator is given by

$$\dot{\hat{z}} = A\hat{z} + Bu + L(y - H\hat{z} - Du) \tag{2.19}$$

$$L = R^{-1}(\gamma_g F E^T + H S)^T \tag{2.20}$$

where $\hat{z}$ is the estimate of the state vector z, L is the Kalman filter gain matrix, S is the solution of the Algebraic Riccati equation using matrix R, and γ_g is the power spectral density of ground acceleration to the sensor noise. In [36], the Kalman–Bucy filter is used as the state estimator represented by

$$\dot{\hat{z}} = A\hat{z} + Bu + L(y - \hat{y}) \tag{2.21}$$

$$L = EC^T R^{-1} \tag{2.22}$$

Kalman filter cannot be applicable if the building parameters; mass, stiffness, and damping are not available, in that case sensors are used for the state estimation. There are different sensors available to measure displacement, velocity, and acceleration [2]. During the seismic excitation, the reference where the displacement and velocity sensors are attached will also move, as a result the absolute value of the above parameters cannot be sensed. Alternatively, accelerometers can provide inexpensive and reliable measurement of the acceleration at strategic points on the structure. A comparative study about the performance of the displacement, velocity, and acceleration sensors are performed in [3] and it is shown that the acceleration sensor is more effective compared to the other two sensors. A number of experiments and implementations about the acceleration feedback in structural control were carried out in [6].

An accelerometer measures the absolute acceleration, which is then integrated for estimating the velocity and displacement. Obtaining the velocity and displacement from the measured acceleration is a practically challenging task. Although time integration of the acceleration seems to be a straightforward solution for estimating the velocity and displacement, there are some practical difficulties that can result in a wrong estimation. Integrating these signals will result in the amplification of low frequencies components, reduction in the magnitude of high frequencies signals, and phase errors. In other words, the low-frequency signals including the DC offset present in the acceleration signal will dominate the result of the velocity and displacement, giving an unrealistic estimation.

The output of the accelerometer $a(t)$ can be expressed as

$$a(t) = k_a \ddot{x}(t) + \varphi(t) + \varepsilon \tag{2.23}$$

where k_a is the accelerometer gain, $\varphi(t)$ is the noise and disturbance effects of the measurement, and ε denotes the DC bias [42]. Accelerometer has different source of noise, integrating these noise signals leads to an output that has a root mean square (RMS) value that increases with integration time, even in the absence of any motion of the accelerometer [50]. The RMS positional error $e_{x(t)}$ of an acceleration signal with a bias ε can be approximated as

$$\text{RMS}\{e_{x(t)}\} = \frac{1}{2}\varepsilon t^2 \tag{2.24}$$

which will grow at a rate of t^2.

It has been shown that the aliasing can cause low-frequency errors in the measured acceleration signal [51]. Aliasing is an unavoidable phenomenon that happens when digitizing the analog signals using an analog-to-digital converter (ADC). During this conversion, the frequency components above the Nyquist rate are folded back into the bandwidth of interest. Then, the acceleration signal in (2.23) can be rewritten as

$$a(t) = k_a \ddot{x}(t) + \varphi(t) + \varepsilon + \ddot{x}_s(t) \tag{2.25}$$

where $\ddot{x}_s(t)$ is the aliasing content due to sampling. This low-frequency content will be amplified during the integration process. This aliasing effect is not completely removable but its effect can be minimized by using an anti-aliasing filter between the accelerometer and data acquisition card. The ADC sampling rate needs to be high enough compared to this filter cutoff frequency and the sampling should to be done in uniform time intervals.

The other source of offset in the measured acceleration is the ADC itself [43]. If the acceleration is slow compared with the quantization level of the conversion, an offset is added into the acceleration signal. This effect can be reduced by increasing the resolution of the ADC.

Apart from these issues, the integration output can also be affected by the integration techniques. The integration methods like the Trapezium rule, Simpson's rule, and Tick's rule have problems with low-frequency components, and they also show instability at high frequencies [52].

A drift-free integrator is proposed by Gavin et al. [53], which is implemented using analog and digital circuits. The paper presents three types of integrators: (1) implemented using a first order low-pass filter as the integrator and two stages of high-pass filters for removing the offset, (2) analog integrator with feedback stabilization, and (3) a stabilized hybrid analog–digital integrator with an exponential accuracy when integrating long-period signals. In another work [54], the drift due to the integration is eliminated by; first filtering the acceleration signal using a frequency-domain filter called Fast Fourier transform-direct digital integration (FFT-DDI) and then is integrated for estimating the velocity and displacement. The same method is repeated for removing the drift occurred due to the unknown initial conditions.

The constant offset present in the acceleration data can be represented using a baseline. The integration may cause a drift in this baseline, which will give a wrong estimation. A baseline correction method is proposed in [55] that uses a least-square curve fitting technique and a frequency-domain filtering for avoiding the drift during the integration. The correction is done by determining a baseline in polynomial form, which is then subtracted from the measured acceleration signal, then is integrated to obtain the velocity and displacement. Finally, a windowed filter is applied to remove the low-frequency noise.

A practical method for calibrating the positional error obtained by double integrating the acceleration signal is discussed in [50]. The double integration of noise using different techniques is also presented. An initial velocity determination method for the displacement estimation from the acceleration data is suggested in [56], which also considers the initial condition in their design. A weighted residual parabolic acceleration time integration method is proposed in [57], where the displacement is assumed to be a fourth-order polynomial, so that the acceleration variation with time is quadratic. A numerical integrator for estimating the velocity and displacement from the measured acceleration signal is proposed in [23]. The effectiveness of the integrator is illustrated experimentally by performing a structural vibration control on a shake table using a PD controller.

2.3 Structural Control Devices

The structural vibration control is aimed to prevent structural damages using vibration control devices. Various control devices have been developed to ensure the safety of the building structure even when excessive vibration amplitudes occur due to earthquake or wind excitations. The control devices are actuators, isolators, and dampers, which are used to attenuate the unwanted vibrations in a structure. Many active and passive devices have been used as vibration control devices. The passive damper modifies the structure response without using an external power supply. Active actuators can generate required forces for controlling the structure dynamics. Using an external power supply, these devices will modify the structure stiffness or damping, which results in a structural dynamics change. The semi-active device combines the properties of both passive and active devices. Hybrid devices are formed either by using both passive and active devices or by using both passive and semi-active devices. Other well-known vibration control devices are the base isolators. The list of the commonly used structural control devices is summarized in the Table 2.2. Basic concepts of some popular devices are discussed below.

Table 2.2 Structural control devices [7, 17, 19]

Passive	Active	Semi-active	Isolator	Hybrid
TMD, TLCD, metallic dampers, friction dampers, viscoelastic dampers, viscous fluid dampers	AMD, active tendons	MR/ER dampers, semi-active TMD, semi-active TLCD, friction control devices, stiffness control devices, viscous fluid dampers	Elastomeric bearings, lead-plug bearings, high-damping rubber bearings, friction pendulum bearings	HMD, HBI

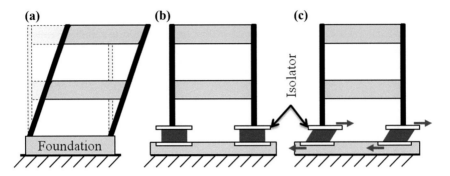

Fig. 2.5 Base isolation system

2.3.1 Base Isolators

Base isolators are flexible isolation devices, placed between the building structure and the foundation for reducing seismic wave propagation into the structure. The addition of this device will increase the flexibility of the structure, hence, the structural time period. For that reason, isolators reduce the propagation of high frequency signal from ground to the structure, which makes it suitable for implementing in small and middle-rise building structures [7]. The Fig. 2.5 shows the changes in the structure response while using base isolator.

Base isolation is one of the popular technique applied widely, especially in the case of bridges. In general, the isolators can be formed using elastomeric bearings, sliding bearings, and combinations of both types of bearings. Elastomeric bearings are made up of elastic materials like the rubber. In the second case, the isolator uses sliding mechanism [7]. In bridges, the isolators are easily implemented by replacing standard bridge bearings by isolation bearings. More information about the types of isolators and their implementation can be found in [58].

Base isolation is well-known passive control technique. But active [59] and semi-active [60] control schemes were also proposed. Another class of base isolation devices is the hybrid base isolation (HBI), made by combining the passive base isolator with the active or semi-active base isolator/control [18]. Sometimes, the seismic activity in the building is reduced by placing isolators between the substructure columns, not in the base, hence called as seismic isolators.

2.3.2 Passive Devices

Structural control using passive devices is called passive control. A passive control device does not require an external power source for its operation and utilizes the motion of the structure to develop the control forces. These devices are normally termed as energy dissipation devices, which are installed on structures to absorb a

Fig. 2.6 Mechanical model
of Building-TMD

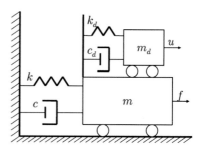

significant amount of the seismic-or wind-induced energy. The energy is dissipated
by producing a relative motion within the control device with respect to the structure motion [19]. For the ideal passive devices, the control forces applied to the
structure are only dependent to the structural motion, which can be mathematically
represented as [6]

$$f_i(t) = -c_i \dot{x}_{di}(t) \tag{2.26}$$

where $\dot{x}_{di}$ is the relative velocity across the ith device and c_i is the damping coefficient
associated with the ith device.

Vibration absorber systems such as tuned mass damper (TMD) has been widely
used for vibration control in mechanical systems. Basically, a TMD is a device
consisting of a mass attached to a building structure such that it oscillates at the same
frequency of the structure, but with a phase shift. The mass is usually attached to the
building through a spring-dashpot system and energy is dissipated by the dashpot as
relative motion develops between the mass and structure [61]. A simple mechanical
model for TMD is depicted in Fig. 2.6. An early study about the TMD with a practical
application is illustrated in [62].

Tuned liquid column damper (TLCD) dissipates energy similar to that of TMD,
where the secondary mass is replaced with a liquid column, which results in a highly
nonlinear response. They dissipate energy by passing the liquid through the orifices.
A simple mechanical model of TLCD is depicted in Fig. 2.7. The natural frequency
of the TLCD can be obtained as [63]

$$\omega_n = \sqrt{\frac{2g}{L_t}} \tag{2.27}$$

where L_t is the length of the liquid tube and g is the acceleration due to gravity.
The equation of motion of a TLCD satisfies the following expression [64]

$$\rho_l \Delta L_t \ddot{x}_v(t) + \frac{1}{2}\rho_l \Delta \xi \, |\dot{x}_v(t)| \, \dot{x}_v(t) + 2\rho_l \Delta g x_v(t) = -\rho_l \Delta L_h \ddot{x}(t) \tag{2.28}$$

where $x_v(t)$ is the vertical displacement, ρ_l is the liquid density, L_h and L_t, respectively are the horizontal and total length of the liquid column, Δ is the area of cross

Fig. 2.7 Mechanical model
of TLCD

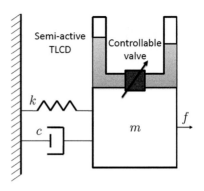

section, and ξ is the headless coefficient. A comparison study of the performance of three types of mass dampers; TMD, TLCD, and Liquid column vibration absorber (LCVA) were discussed in [59] and it is concluded that the TMD performs better than the other two dampers.

Other passive dampers are [17]: metallic yield dampers which dissipate the energy through the inelastic deformation of metals, friction dampers which utilize the mechanism of solid friction, develops between two solid bodies sliding relative to one another, to provide the desired energy dissipation, and viscoelastic dampers that dissipates the energy through the shear deformation.

Viscous fluid damper works based on the concept of sticky consistency between the solid and liquid. It has a movable piston within a housing filled with highly viscous fluid. The piston contains a number of orifices, through which the fluid passes from one side to another that will result in energy dissipation. The output force of the orifice controlled viscous fluid devices can be expressed as [65]

$$f(t) = c\,|\dot{x}_d(t)|^{\alpha_c}\ \text{sgn}(\dot{x}_d(t)) \tag{2.29}$$

where $\dot{x}_d$ is the relative velocity of the viscous fluid device and α_c is a coefficient in the range of 0.3–2.0.

Passive dampers are very simple and due to the fact that it will not add energy to the structure, hence it cannot make the structure unstable. Most of the passive dampers can be tuned only to a particular structural frequency and damping characteristics. Sometimes, these tuned values will not match with the input excitation and the corresponding structure response. For example; (1) nonlinearities. in the structure cause variations in its natural frequencies and mode shapes during large excitation, (2) a structure with a multiple-degree-of-freedom (MDOF) moves in many frequencies during the seismic events. As the passive dampers cannot adapt to these structure response changes, it cannot assure a successful vibration suppression [9]. This is the major disadvantage of the passive dampers, which can be overcome by using multiple passive dampers, each tuned to different frequencies (e.g., doubly TMD, Multiple TMD) or by adding an active control to it.

2.3.3 Active Devices

The concept of active control has started in early 1970 s and the full-scale application was performed in 1989 [18]. An active control system can be defined as a system that typically requires a large power source for the operation of electrohydraulic or electromechanical (servo motor) actuator, which increases the structural damping or stiffness. The active control system uses sensors for measuring both the excitation and structural responses, and actuators for controlling the unwanted vibrations [19]. The working principle of the active control system is that, based on the measured structural response the control algorithm will generate control signal required to attenuate the vibration. Based on this control signal, the actuators placed in desired locations of the structure generate a secondary vibrational response, which reduces the overall structure response [66]. Depending on the size of the building structure, the power requirements of these actuators vary from kilowatts to several megawatts [67]. Hence, an actuator capable of generating a required control force should be used. As the active devices can work with a number of vibration modes, it is a perfect choice for the MDOF structures. A number of reviews on active structural control were presented [11].

The ideal actuators are assumed to have the ability to instantaneously and precisely supply force commanded by the control algorithm [6]. There are many active control devices designed for structural control applications. A recent survey on active control devices is presented in [9]. An active mass damper (AMD) or active tuned mass damper (ATMD) is created by adding an active control mechanism into the classic TMD. In this system, 1 % of the total building mass is directly excited by an actuator with no spring and dashpot attached. ATMD control devices were first introduced in [68]. These devices are initially used to reduce structural vibrations under strong winds and moderate earthquake.

Active tendons are prestressed cables, where its stress is controlled using actuators for suppressing the vibration [9]. The structural vibration control using active cables and tendons is presented in [12]. Various numerical analytical studies have been carried out using tendons for active control [69]. At low excitations, the active control system can be switched-off, then the tendons will resist the structural deformation in passive mode. At higher excitations, active mode is switched-on to reach the required tension in tendons.

A comparison study between active and passive control systems was carried out in [6] using H_2/LQG control algorithm. In simulation, it is found that for SDOF structure both the active and passive control systems performed similarly, whereas in the case of structure with MDOF the active control system showed high performance.

The active control devices found to be very effective in reducing the structural response due to high magnitude earthquakes. However, there are some challenges left to the engineers, such as how to eliminate the high power requirements, how to reduce the cost, and maintenance. These challenges resulted in the development of semi-active and hybrid control devices [70].

2.3.4   Semi-active Devices

A semi-active control system typically requires a small external power source for its operation and utilizes the motion of the structure to develop control force, where the magnitude of the force can be adjusted by an external power source [19]. It uses the advantages of both active and passive devices. The semi-active devices for structural control application were first proposed by Hrovat et al. in 1983 [71].

The benefits of the semi-active devices over active devices are their less power requirements, which can even be powered using a battery that is more important during the seismic events, when the main power source to the building may fail. Semi-active devices cannot inject mechanical energy into the controlled structural system, but has properties that can be controlled to optimally reduce the response of the system. Therefore, in contrast to active control devices, semi-active control devices do not have the potential to destabilize (in bounded-input bounded-output sense) the structural system [70]. A detailed review of semi-active control systems is provided in [72].

Like passive friction dampers, these semi-active frictional control devices dissipate energy through friction caused by the sliding between two surfaces. For this damper, a pneumatic actuator is provided in order to adjust the clamping force [73]. An ideal friction damper can be modeled as a Coulomb element, where the output force is termed as

$$f = \mu f_n \, \text{sgn}(\dot{x}) \qquad\qquad (2.30)$$

where μ is the friction coefficient and f_n is the normal force [19]. In the case of friction dampers, the friction coefficient needs to be tuned to have a good energy dissipation. In contrast with the passive friction dampers, the semi-active friction dampers can easily adapt the friction coefficient to varying excitations from weak to strong earthquakes.

Semi-active controllable fluid dampers are one of the most commonly used semi-active control device. For these devices, the piston is the only moving part, which makes them more reliable. These devices have some special fluid, where its property is modified by applying external energy field. The electric and magnetic fields are mainly used to control these devices, which is so-called as Electro rheological (ER) and magneto rheological (MR) dampers, respectively [17].

ER damper [19]: ER dampers consist of liquid with micron-sized dielectric particles within a hydraulic cylinder. When an electric field is applied, these particles will polarize due to the aligning, thus offers more resistance to flow resulting a solid behavior. This property is used to modify the dynamics of the structure to which it is attached.

MR damper [19]: The construction and functioning of MR dampers are analogous to that of ER dampers, except the fact that instead of the electric field, magnetic field is used for controlling the magnetically polarizable fluid. MR dampers have many advantages over ER dampers, which made them more popular in structural control applications. These devices are able to have a much more yield stress than ER with less

Table 2.3 MR and ER damper properties

Property	MR damper	ER damper
Max. yield stress	50–100 kPa	2–5 kPa
Maximum field	~250 kA/m	~4 kV/mm
Plastic viscosity	0.1–1.0 Pa-s	0.1–1.0 Pa-s
Operable temperature range	−40 to 150 °C	+10–90 °C
Stability	Unaffected by most impurities	Cannot tolerate impurities
Response time	milliseconds	milliseconds
Density	3–4 g/cm^3	1–2 g/cm^3
Maximum energy density	0.1 J/cm^3	0.001 J/cm^3
Power supply (typical)	2–25 V; 1–2 A	2000–5000 V; 1–10 mA

input power. Moreover, these devices are less sensitive to impurities. A comparison between MR and ER fluid dampers are summarized in Table 2.3.

Different modeling techniques are available to express the behavior of these devices, such as; Bingham model, Bingham viscoplastic model, Gamota and Filisko model, Bouc–Wen model, modified Bouc–Wen model, etc. [74]. Among these techniques, Bingham model is the simplest modeling tool for both ER and MR dampers. When any field is applied to these devices, the change in the fluid property can be modeled using a Bingham viscoplastic model [12]. The plastic viscosity of this model is given in terms of the shear stress and shear strain, which is mathematically represented as

$$\tau = \tau_y \, \mathrm{sgn}(\dot{\gamma}) + \eta \dot{\gamma} \qquad (2.31)$$

where τ is the total shear stress, τ_y is the yield stress due to the applied field, $\dot{\gamma}$ is the rate of the shear strain, and η is the plastic viscosity. The relationship between the force and displacement of a MR damper using this model is given by [75]

$$f = \frac{12\eta_N L_p \Delta_p^2}{\pi D_i D_p^3} \dot{x}(t) + \frac{3 L_p \tau_y}{D_p} \Delta_p \, \mathrm{sgn} \, [\dot{x}(t)] \qquad (2.32)$$

where L_p is the piston length, Δ_p is the piston cross-sectional area, D_i is the inner diameter, D_p is the diameter of the small gap in the piston, and η_N is the Newtonian viscosity independent of the applied magnetic field. The yield stress can be represented as a function of the control current I as follows.

$$\tau_y = A_1 e^{-I} + A_2 \ln(I + e) + A_3 I \qquad (2.33)$$

where A_1, A_2 and A_3 are the coefficients relative to the MR fluid property and e is the Euler's number.

Bingham model is a mechanical version of the Bingham viscoplastic model, which uses damping and Coulomb friction components in the model. This model is further extended, known as Gamota and Filisko model, which is a parametric viscoelastic-plastic model. But all of these methods have some shortcomings especially at low velocities. The classic Bouc–Wen can model the hysteresis loop pretty well, but fails to predict the roll-off problem seen at low velocities. A modified Bouc–Wen model was proposed by Spencer et al. [74], where an additional damping (c_1) and stiffness (k_1) is added to compensate the roll-off and accumulator stiffness, respectively. The total force of the MR damper is obtained as

$$
\begin{aligned}
f &= \alpha_b \tilde{z} + c_0(\dot{x} - \dot{\tilde{y}}) + k_0(x - \tilde{y}) + k_1(x - x_0) \\
  &= c_1 \dot{\tilde{y}} + k_1(x - x_0)
\end{aligned}
\tag{2.34}
$$

where $\tilde{y}$ and $\tilde{z}$ can be found as

$$
\dot{\tilde{z}} = -\gamma_m \left| \dot{x} - \dot{\tilde{y}} \right| \tilde{z} |\tilde{z}|^{\tilde{n}-1} - \beta_m \left(\dot{x} - \dot{\tilde{y}} \right) |\tilde{z}|^{\tilde{n}} + \delta_m \left(\dot{x} - \dot{\tilde{y}} \right)
\tag{2.35}
$$

$$
\dot{\tilde{y}} = \frac{1}{c_0 + c_1} \{ \alpha_b \tilde{z} + c_0 \dot{x} + k_0(x - \tilde{y}) \}
\tag{2.36}
$$

where c_0 is the viscous damping at large velocities, c_1 is the viscous damping for force roll-off at low velocities, k_0 is the stiffness at large velocities, k_1 is the damper accumulator stiffness, and x_0 is the initial displacement of spring. k_1 and α_b is a third-order polynomial. The corresponding mechanical model is depicted in Fig. 2.8. In [55], the dynamic modeling and two quasi-static models (axisymmetric and parallel-plate model) of the MR damper are studied through experiments.

The semi-active fluid viscous damper consists of a hydraulic cylinder, which is separated using a piston head. The cylinder is filled with a viscous fluid, which can pass through the small orifices. An external valve which connects the two sides of the cylinder is used to control the device operation. The semi-active stiffness control device modifies the system dynamics by changing the structural stiffness [19].

Fig. 2.8 Modified Bouc–Wen model of MR damper

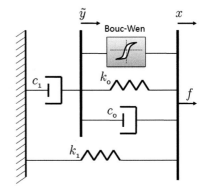

2.3.5 Hybrid Devices

Hybrid actuators combine robustness of the passive device and high performance of the active devices. Due to the inclusion of multiple control devices, the hybrid system overcomes the limitations and restrictions seen in the single control devices like passive, active, and semi-active devices. The hybrid systems are further classified into two classes: HBI and hybrid mass damper (HMD) [18]. As the base isolation exhibits nonlinear behavior, various nonlinear control technologies like the robust control were adopted to control these hybrid devices [70].

HMD can be formed by combining the passive devices like TMD along with some active devices like AMD. The capability of the TMD is increased by adding a controlling actuator to it, which increases the system robustness in changing the structure dynamics. These HMDs are found to be cost effective in terms of the energy requirement for their operation, when compared with active control systems [18]. The full-scale implementation of active structural control systems in Japan, USA, Taiwan, and China are enlisted in [17], where the HMD is found to be the most commonly employed device compared with other devices.

Researchers have also investigated the various control methods for HMD, like optimal control methods, sliding mode control, gain scheduling, etc. [70]. As these systems utilize two types of actuators, it will have a series of objective functions, which results in a multi-objective optimization problem. To derive an optimal solution, a preference-based optimization model using GA is proposed. The designed model is compared with a non-hybrid system and is found to be very cost effective in suppressing the vibrations. A hybrid system using the HMD and a viscous damper is discussed in [76] for the reduction of the wind-induced vibrations of high-rise building.

The implementation of the above-mentioned devices will result in different control schemes, which are summarized in Fig. 2.9. In the passive control, the passive device reduces the vibration response of a structure without using any feedback, see Fig. 2.9b. In the active and semi-active case, the input and output response of the structure is measured and based on that the controller generates a desirable output command signal. This signal is then used to drive the active or semi-active devices for attenuating the vibration, which are shown in Fig. 2.9c, d, respectively. In the case of hybrid control shown in Fig. 2.9e, only the active/semi-active device uses the feedback, whereas the passive devices works independently.

Typical installations of control devices are shown in Fig. 2.10. Other recent technique is the connected control method, where the adjacent buildings are interconnected using control devices for vibration attenuation, see Fig. 2.11. In [77], passive devices are installed between the adjacent structures for inter-structure protection and at the same time semi-active dampers are placed in the building floors for protecting the substructure.

A brief state-of-the-art review about the structural control devices can be found in [17]. The simplicity of the passive systems made them more common in seismic control applications. The active systems including the semi-active and hybrid systems,

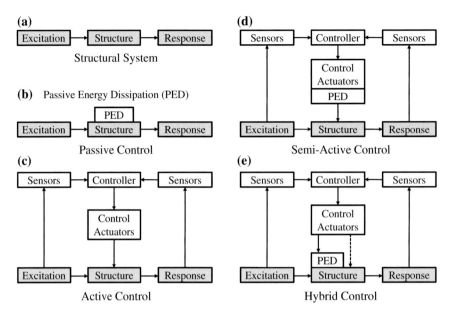

Fig. 2.9 Control schemes [17]

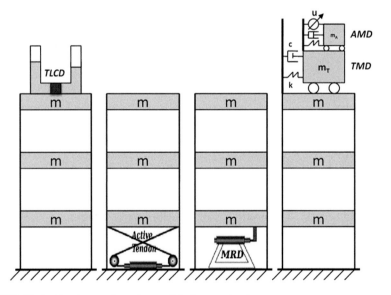

Fig. 2.10 Typical implementation of control devices on structures

generates a control force based on the measurements of the structural responses. Due
to this ability of measuring the structural response it can be designed to accommodate
a variety of disturbances, which makes them to perform better than the passive systems. More on the governing equations of dampers and actuators can be found in [47].

Fig. 2.11 Buildings
interconnected using
dampers

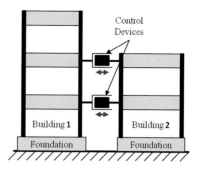

Control devices are used to control the dynamics of the structure to a desired response. Therefore, the dynamic model of a structure will change once a control device is installed on it. That is, it is expected that the installation of a control device will modify the structure parameters like its natural frequency, thereby changing the system model [36]. As a consequence, it is necessary to consider the dynamics of the actuator in the structure.

Consider a passive damper added to a structure represented in (2.37), then the system model can be rewritten as [17]

$$m\ddot{x} + c\dot{x} + kx + \Pi(x) = -(m + m_d)\ddot{x}_g \tag{2.37}$$

where m_d is the mass of the damper and $\Pi(x)$ represents the force corresponding to the damper, used to modify the structure response for reducing vibrations. The same formulation can be done in the case of active control devices, where (2.37) can be rewritten as follows

$$m\ddot{x} + c\dot{x} + kx = -mu(t) - m\ddot{x}_g \tag{2.38}$$

If the control force is selected as per the relationship given in (2.39)

$$u(t) = \frac{\Pi(x)}{m} \tag{2.39}$$

then (2.38) becomes

$$m\ddot{x} + c\dot{x} + kx + \Pi(x) = -m\ddot{x}_g \tag{2.40}$$

In contrast to the passive control method, here, the control function $\Pi(x)$ is derived as a control law.

The motion equation of a structural system with n-DOF and o control devices subjected to an earthquake excitation can be expressed as

$$M\ddot{x}(t) + C\dot{x}(t) + Kx(t) = \Gamma u(t) - M\Lambda\ddot{x}_g(t) \tag{2.41}$$

where $u(t) \in \mathbb{R}^{n \times 1}$ is the control force vector and $\Gamma \in \mathbb{R}^{n \times o}$ is the location matrix of the control devices. Equation (2.41) becomes nonlinear if the control force is generated using a nonlinear device, such as MR damper or by using a nonlinear control algorithm, such as intelligent control.

2.4 Active Structural Control Techniques

The objective of structural control system is to reduce the vibration and to enhance the lateral integrity of the building due to earthquakes or large winds, through an external control force [21]. In active control system, it is essential to design one controller in order to send an appropriate control signal to the control devices so that it can reduce the structural vibration. The control strategy should be simple, robust, fault tolerant, need not be an optimal, and of course must be realizable [22].

2.4.1 Linear Control of Building Structures

PID Control. The proportional-integral-derivative (PID) has been widely conducted for practical applications, especially for the systems with one or two DOF. For multivariable systems, its control algorithm becomes more complex, which makes them unsuitable for the applications like vibration control of MDOF flexible structures. A simulation was carried out for a simple proportional controller, which is able to reduce the building displacement for wind excitation, but found to be ineffective for strong earthquake excitation [13].

In [78], two PD controllers were used for controlling two actuators installed in the first and fifteenth floor. The control law is given as

$$u(t) = K_p \left[e(t) + K_d \frac{de(t)}{dt} \right] \tag{2.42}$$

where K_p and K_d are the proportionality constant and derivative time, respectively, and $e(t)$ is the position error. The designed PD controller performance is found to be less efficient when compared with that of a fuzzy logic controller (FLC).

In a work done by [79], a PID controller is designed which have the following controlling law

$$u(t) = K_p \left[e(t) + \frac{1}{K_i} \int_0^t e(t)dt + K_d \frac{de(t)}{dt} \right] \tag{2.43}$$

where K_i is the integral gain. Here, the PID performance is compared with that of a sliding mode controller (SMC) and found to be less effective in controlling the structural vibration. In [80], a proportional-integral (PI) controller is used to actuate the AMD against the structural motion due to earthquake.

H$_\infty$ Control. H$_\infty$ technique is one of the widely used linear robust control scheme in structural vibration control. This technique is insensitive with respect to the disturbances and parametric variations, which makes them suitable for the MIMO type structural control systems [81].

A modified H$_\infty$ controllers, for example, pole-placement H$_\infty$ control is presented in [76]. In this work, instead of changing the structure stiffness some target damping ratio is considered. A bilinear transform is adopted to locate the closed-loop poles in a specific region within the H$_\infty$ controller design framework. The relation between the final closed-loop poles and bilinear transform parameters is derived as a quadratic equation and using this equation a new noniterative direct method is developed for an optimal H$_\infty$ controller design.

Normally, the H$_\infty$ design results in a higher order system, which will make the implementation more difficult [62]. So it may be necessary to reduce its order, which can be done by performing balanced truncation. The truncation has two classes; direct method and indirect method. The balanced truncation assures very few information losses about the system, which is achieved by truncating only less controllable and observable states. It is shown that the performance of the reduced low order system is nearly same as the performance of the actual higher order controller.

A H$_\infty$ based structural controller using Takagi–Sugeno Fuzzy model was proposed in [82]. The controller stability is derived based on Lyapunov stability theory, which is evaluated as a LMI problem. If the initial condition is considered, the H$_\infty$ control performance satisfies the following condition:

$$\int_0^{t_f} z^T(t) Q z(t) dt \leq z^T(0) P z(0) + \epsilon^2 \int_0^{t_f} \ddot{x}_g^T \ddot{x}_g dt \qquad (2.44)$$

where t_f denotes the termination time of the control, P and Q are the positive definite matrices, and ϵ denotes the effect of $\ddot{x}_g$ on $z(t)$. The effectiveness of the proposed algorithm is demonstrated through numerical simulations on a 4-story building.

As discussed earlier, time-delay is an important factor to be considered while designing a control system. A H$_\infty$ controller is presented in [83], which considers time-delay in control input u. The proposed algorithm determines the feedback control gain with a random search capability of GA and solving a set of LMIs. The effectiveness of the proposed algorithm is proved through simulation of a system with larger input time-delay.

Optimal Control. Optimal control algorithms are based on the minimization of a quadratic performance index termed as cost function, while maintaining a desired system state and minimizing the control effort [13]. The most basic and commonly used optimal controller is the linear quadratic regulator (LQR). For structural control applications, the acceptable range of structure displacement and acceleration are considered as the cost function that is to be minimized.

An energy-based LQR is proposed in [69], where the controller gain matrix is obtained by considering the energy of the structure. The structural energy is defined as

$$\frac{1}{2}\dot{x}^T(t)M\dot{x}(t) + \frac{1}{2}x^T(t)Kx(t) \qquad (2.45)$$

where the first term is the kinetic energy and the second term is the potential energy of the structural system.

A modified LQR is proposed in [4], which is formed by adding an integral and a feedforward control to the classic LQR. A state feedback gain and an integral gain are used to reduce the steady-state error. A feedforward control is included to suppress the structural responses and to reduce the effect of earthquakes. A structural vibration control utilizing a filtered LQ control is presented in [84]. As all the structural state variables are not observable, a suboptimal control is used, where the system states are reduced using low-pass filters. A LQR based on GA is presented [24], where the GA is used for choosing the weighting matrix.

Sometime, states of the structures are measured indirectly using some observers like Kalman filters. The addition of a Kalman filter to a LQR control strategy leads to what is termed as Linear Quadratic Gaussian (LQG) [36]. In other words, LQG is formed by combining the linear quadratic estimator with LQR. These LQG are generally used for the systems which has Gaussian white noise [10].

The conventional LQG controller sometimes do not consider the input force term in their design. Chen [82] proposed an active vibration control scheme using a combination of LQG and an input estimation approach. The input estimation approach is introduced to observe the input disturbance forces for the open loop control, that is used to cancel out the input forces. The proposed method is evaluated through numerical experiments on linear lumped-mass systems and a better performance is reported compared to that of the conventional LQG.

An active controller utilizing MR damper is designed using LQG control strategy under a wind loading by means of drag forces [36]. A real set of recorded wind speed data is used to excite the laboratory prototype. A H_2/LQG based controller is presented in [33], which uses wireless sensing motes (MICA2) for sensing the acceleration signal. More works about the structural control using LQR/LQG control algorithms can be found in [85]. Optimal algorithms based on instantaneous optimal control has also been developed for nonlinear systems. The nonlinear optimal methods using GA, FLC, etc., will be discussed later.

2.4.2   Intelligent Control of Building Structures

Neural Network Control. In recent years, the structural control systems based on NN are very popular, because of its massively parallel nature, ability to learn, and its potential in providing solutions to the foregoing unsolved problems. They provide a general framework for modeling and control of nonlinear systems such as building structures.

In the middle of the 1990s, very few structural control applications have been reported based on NN. Wen et al. [86] presented a NN-based active control of a

SDOF system that can become nonlinear and inelastic. One inverse mapping NN and one emulator NN are used in the design. The difference between the actual overall structural response and response due to the control force only, is used as the input to the inverse mapping NN. The emulator NN predicts the response of the structural system to the applied control force. Using this response, a control force with a phase shift is generated to nullify the excitation.

A backpropagation (BP)-based ANN for active control of SDOF structure is proposed by Tang [22]. This control strategy does not need the information of the external excitation in advance and the control force needed for the next sampling time is completely determined from the currently available information. The ANN with five neuron elements (displacement, velocity, and load of the preceding time step and displacement and velocity of the current time step) is used, which will perform two sequential calculations in every sampling interval; (a) calculate the load (b) based on the calculated load, the control force $u(t)$ needed for the next time interval is calculated. Apart from the numerical verification of the above algorithm, they have also presented a study on the uncertainties in the system modeling and input motion.

Consider the minimization of the cost function in a discrete form with total time step r and increment time Δt

$$\hat{J} = \sum_{\hat{n}=0}^{r} \hat{J}_{\hat{n}} = \frac{1}{2} \sum_{\hat{n}=0}^{r} (z[\hat{n}]^T Q z[\hat{n}] + u[\hat{n}]^T R u[\hat{n}]) \Delta t \qquad (2.46)$$

where $\hat{J}_{\hat{n}}$ is the instantaneous cost function, $\hat{J}$ is the global cost function, and $\hat{n}$ is the discrete time steps. If the weights are updated at each time step in order to minimize the instantaneous cost function, this learning mode is called pattern learning, and if the weights are updated once for all time steps so that the global cost function $\hat{J}$ is reduced, this learning mode is known as batch learning. An optimal control algorithm using NN based on the pattern learning mode is presented in [63]. The steepest-descent method is used here as the weight updating rule.

One multilayer NN controller with a single hidden layer is presented in [87]. The optimal number of hidden neurons is selected after performing a number of iterative training cycles. The network will generate an active control force as output using the structure response as its input. The batch learning is used here, where the network weights and biases are selected in such a way that a minimal objective cost function is achieved. The steepest-gradient-descent optimization method is used for the weight update, where the partial-differential equations are solved using the chain rule.

Probabilistic neural networks (PNN) are feed-forward networks built with three layers. They are derived from Bayes decision networks that estimates the probability density function for each class based on the training samples. The PNN trains immediately but execution time is slow and it requires a large amount of memory space. A new method to prepare the training pattern and to calculate PNN output (control force) quickly is proposed in [88]. The training patterns are uniformly distributed at the lattice point in state-space, so that the position of invoked input can be known. This type of network is called as Lattice probabilistic neural network (LPNN). The calculation time is reduced by considering only the adjacent patterns. Here, the distance between the input pattern (response of structure) and training patterns (lattice type) for LPNN are calculated, which is then converted as the weights.

An active type NN controller using one counterpropagation network (CPN) is presented in [89], which is an unsupervised learning type NN, so that the control force is generated without any target control forces. Another intelligent control technique using a NN is proposed for seismic protection of offshore structures [65].

The ability of the nets to perform nonlinear mappings between the inputs and outputs, and to adapt its parameters so as to minimize an error criterion, make the use of ANN particularly well suited for the identification of both linear and nonlinear dynamic systems. The NN for system identification in structural control applications were presented in [40]. A NN is designed to approximate the nonlinear structural system and the corresponding stability conditions are derived [82]. A state-feedback controller for the NN is designed using a linear differential inclusion (LDI) state-space representation, which is useful in the stability analysis. Using NN, the system in (2.14) is approximated as a LDI representation with less modeling errors.

An intelligent structural control system with improved BP-NN is proposed in [90], which is used to predict the inverse model of the MR damper and for eliminating time-delay in the system. The system represented in (2.14) is considered here. The system has two controllers; the first one modifies the actual structural model, which was offline trained before and the second controller causes error emendation by means of online feedback. A multilayer NN for structural identification and prediction of the earthquake input is presented in [91].

Fuzzy Logic Control. Like NN, Fuzzy logic is also a model free approach for system identification and control. The FLC design involves; the selection of the input, output variables, and data manipulation method, membership function, and rule base design. Due to its simplicity, nonlinear mapping capability, and robustness, the FLC has been used in many structural control applications [92].

A FLC is designed [78] for a 15-story structure with two type of actuators, one mounted on the first floor and the other actuator (ATMD) on the fifteenth floor. The proposed FLC uses the position error and their derivatives as the input variable to produce the control forces for each actuator. The rule base is formed using seven fuzzy variables. The controller uses Mamdani method for fuzzification and Centroid method for defuzzification. A simulation using Kocaeli earthquake signal is carried out to prove the improvement in the performance of the FLC. A similar type of FLC is presented in [93], for the active control of wind excited tall buildings using ATMD. Another FLC for MDOF is proposed [94], that uses the same architecture, which is further modified into MDOF using weighted displacement and weighted velocity. In order to get the maximum displacement and velocity values, a high magnitude earthquake is used to excite the building structure. As all the floors do not have control devices, a weighting value is assigned to each floor, which will be large if the control device is closer to that particular floor. Finally, a force factor is calculated based on the weights of each floor.

A Fuzzy based on-off controller is designed to control the structural vibration using a semi-active TLCD [95]. The optimal control force is given as

$$u = -\sum_{i=1}^{r} p_i z_i \qquad (2.47)$$

where $p_i = [p_1, ..., p_r]$ is the optimal control gain vector obtained using LQR strategy. The control force will act opposite to the direction of the liquid velocity $(\dot{x}_f)$. The regulation of the control force is done by varying the coefficient of headloss (ξ) with the semi-active control rule as given below.

$$\xi(t) = \begin{array}{ll} \xi_{max} & \text{if} \quad \{z_l(t)\dot{x}_f(t)\} < 0 \\ \xi_{min} & \text{if} \quad \{z_l(t)\dot{x}_f(t)\} \geq 0 \end{array} \qquad (2.48)$$

where z_l represents the largest weighted state, which contributes most of the control force in (2.47). Finally, using the above control law a FLC is designed, that takes the liquid velocity and the large weighted displacement $(z_i = z_l)$ as its input and produces the coefficient of headloss as the output, which is used to control the valve in the semi-active TLCD.

A fuzzy supervisory control method is presented in [96], which has a fuzzy supervisor in the higher level and three subcontrollers in the lower level. First, the subcontrollers are designed based on the LQR strategy, where the three subcontrollers are derived from three different weight matrices. The fuzzy-supervisor tunes these subcontrollers according to the structure's current behavior. A similar work is done in [56], where the subcontroller is designed using an optimal controller in the modal space. The matrix in the Riccati equation is calculated using the natural frequencies of the dominant modes and a corresponding gain matrix is determined. Another FLC for active control of structure using modal space is presented in [97], which uses a Kalman filter as an observer for the modal state estimation and a low-pass filter for eliminating the spillover problem.

Instead of using a mathematical model, a black-box based controller is proposed in [98]. Here, the force-velocity characteristics of the MR damper corresponding to different voltages are obtained experimentally, which are used to calculate the desired control force. The effect of the damper position and capacity on the control response is also studied.

An alternative to the conventional FLC, using an algebraic method is proposed in [99]. Here, the hedge algebra is used to model the linguistic domains and variables and their semantic structure is obtained. Instead of performing fuzzification and defuzzification, more simple methods are adopted, ermed as semantization and desemantization, respectively. The hedge algebra-based fuzzy system is a new topic, which was first applied to fuzzy control in 2008. Compared to the classic FLC, this method is simple, effective, and can be easily interpreted.

Some structural vibration controllers were designed, where the FLC is combined with the GA [5]. The GA is known for its optimization capabilities. The GA is used here to optimize different parameters in the FLC like its rule base and membership function.

Genetic Algorithm. The GA is an iterative and stochastic process that proceeds by creating successive generation of offsprings from parents by performing the operations like selection, crossover, and mutation. The above operation is performed based on the fitness (termed as cost function in optimization problems) value assigned to each individual. After these operations, the parents are replaced by the offsprings, which is continued till an optimal solution for the problem is attained [100].

The structural control problem consists of different objectives to be optimized, which can be formulated using multi-objective optimization algorithms like GA. In [101], a preference-based optimum design using GA for an active control of structure is proposed, where the structure and control system is treated as a combined system. Here, the structural sizing variables, locations of actuators, and the elements of the feedback gain matrix are considered as the design variables and the cost of structural members, required control efforts, and dynamic responses due to earthquakes are considered as the objective functions to be minimized. For each objective criterion, preference functions are defined in terms of degrees of desirability and regions that represent the degrees of desirability. They are categorized as desirable, acceptable, undesirable, and unacceptable with ranges defined by $(\lambda_i \le c_{i_1})$, $(c_{i_1} \le \lambda_i \le c_{i_2})$, $(c_{i_2} \le \lambda_i \le c_{i_3})$, and $(c_{i_3} \le \lambda_i)$ respectively, where c_{i_1}, c_{i_2}, and c_{i_3} are the range boundary values and λ_i is the i-th design objective. The preference-based optimization problem model is then given as

$$F_P(\lambda) = \frac{1}{l} \sum_{i=1}^{l} f_{P_i}[\lambda_i(d)] \qquad (2.49)$$

with $\lambda_i(d) < c_{i_3}$ and $d_{min} \le d \le d_{max}$, where F_P is the aggregate preference function, f_P is the power function, l is the number of design objectives, and d is the vector of design variables; d_{min} and d_{max} are the prescribed design constraints, respectively. Finally, the fitness function of n_g randomly created strings is defined as follows

$$F_{f_i} = [\max(F_{P_j}) + \min(F_{P_j})] - F_{P_i} \quad j = 1, ..., n_g \qquad (2.50)$$

where F_{f_i} is the fitness value of ith individual. A numerical simulation of an earthquake excited 10-story building is carried out and the proposed algorithm is able to achieve improved performance with less control effort.

An active control of structures under wind excitations using a multilevel optimal design based on GA is proposed [25]. The proposed multilevel genetic algorithm (MLGA) considers the number and position of the actuators and control algorithm as multiple optimization problems. This problem has the properties of nonlinearity, noncontinuous, and multimodal objective function. In [102], a GA is used to tune the mass, damping, and stiffness of the MRF absorber.

In [18], a feedback controller is designed, where the feedback gains are optimized using a GA. The controller also considers the time-delay in applying control forces to the devices. Two objective functions are: (1) to reduce the displacement and acceleration response of the ith floor, and (2) to reduce the story drift response as shown in (2.51) and (2.52) respectively.

$$\alpha_1 \sum_{\hat{n}=1}^{r} x_i[\hat{n}]^2 + \alpha_2 \sum_{\hat{n}=1}^{r} \ddot{x}_i[\hat{n}]^2 \qquad (2.51)$$

$$\max \left\{ \sum_{\hat{n}=1}^{r} \frac{|d_1[\hat{n}]|}{x_{m0}}, \sum_{\hat{n}=1}^{r} \frac{|d_2[\hat{n}]|}{x_{m0}}, ..., \sum_{\hat{n}=1}^{r} \frac{|d_m[\hat{n}]|}{x_{m0}} \right\} \qquad (2.52)$$

where α_1 and α_2 are the weights of displacement and acceleration responses respectively, $d_i[\hat{n}]$ is the story drift from the ith to $(i-1)$th floor at the time data point $\hat{n}$, and x_{m0} is the maximum displacement responses in all stories. The effectiveness of the proposed method is demonstrated using numerical simulation of a 3-story and 8-story structures excited by different seismic forces.

The disadvantage of the GA is that, it requires long computational time if the number of variables involved in the computation increases. A modified GA strategy is proposed in [41] to improve the computational time efficiency, which uses the search space reduction method (SSRM) using a Modified GA based on migration and artificial selection (MGAMAS) strategy. In order to improve the computational performance, the algorithm utilizes some novel ideas including nonlinear cyclic mutation, tagging, and reduced data input

Sliding Mode Control. SMC is one of the most popular robust control techniques. A switching control law is used to drive the system's state trajectory onto a prespecified surface in the state-space and to maintain the system's state trajectory on this surface for subsequent time, which results in a globally asymptotically stable system. In the case of structural vibration control, this surface corresponds to a desired system dynamics. The robustness of the SMC against the uncertainties and parameter variations makes them a better choice for structural control applications.

The nonlinear control force in SMC is given as

$$u = u_{eq} - \eta \, \text{sgn}(\sigma(t)) \qquad (2.53)$$

where the linear term u_{eq} is the equivalent control force, $\sigma = [\sigma_1, ..., \sigma_n]$ are the n sliding variables, and η is the design parameter that guarantees the system trajectories reach the sliding surface in finite time. A SMC with hybrid control is proposed in [103], where the control law also termed as reaching law is formed using the constant plus proportional rate reaching law and power rate reaching law.

Due to the imperfection in the high-frequency discontinuous switching, the direct implementation of the control given in (2.53) will result in chattering effect, which may cause damage to the mechanical components, hence the actuators. This effect should be eliminated by suitably smoothing the control force or by using continuous SMC. Many structural control strategies based on the non-chattering SMC were reported [79].

A modal space sliding mode control (MS-SMC) method is designed in [44], where the dominant frequencies are derived using power spectrum as well as the wavelet analysis of the time series of the input-output. SMC based on a single-mode (first mode) reduced-order model is designed. Another SMC based on the modal analysis is presented in [104], where the first six modes of the structure were considered.

During seismic events, the main control unit may lose its functionality, so it is a better option to use a decentralized system, where the whole control is divided into subsystems and are controlled independently. Such a type of decentralized system with SMC is presented in [105]. The numerical studies were carried out for full control and partial control cases and reaching laws were derived for cases; with and without considering actuator saturations. They found that the full control case is more effective, and they could not find any significant changes in the control for different subsystem configurations.

A NN-based SMC for the active control of seismicity excited building structures is proposed in [106]. Here apart from the sliding variables, the matrix σ also represents the slope of the sliding surface. This slope moves in a stable region, which results in a moving sliding surface. A four layer feedforward NN is used to reduce chattering effect and to determine the sliding surface slope. To achieve a minimum performance index, the controller is optimized using a GA during the training process. It is shown that a high performing controller is achieved by using the moving sliding surface. Another SMC based on radial basis function (RBF) NN is reported in [107]. The chattering free SMC is obtained using a two-layered RBF-NN. The relative displacement of each floor is fed as the input to the NN and the design parameter η is taken as the output. A modified gradient-descent method is used for updating the weights.

Couple of research works were carried out in designing the SMC using Fuzzy logic so-called, fuzzy sliding mode control (FSMC) [108]. The SMC provides a stable and fast system, whereas the FLC provides the ability to handle a nonlinear system. The Chattering problem is avoided in most of these FSMC systems. A FSMC based on GA is presented in [109], where the GA is used to find the optimal rules and membership functions for the FLC.

References

1. J.T.P. Yao, Concept of structural control. J. Struct. Div. **98**, 1567–1574 (1972)
2. G.W. Housner, L.A. Bergman, T.K. Caughey, A.G. Chassiakos, R.O. Claus, S.F. Masri, R.E. Skelton, T.T. Soong, B.F. Spencer, J.T.P. Yao, Structural control: past, present and future. J. Eng. Mech. **123**, 897–974 (1997)
3. T. Balendra, C.M. Wang, N. Yan, Control of wind-excited towers by active tuned liquid column damper. Eng. Struct. **23**, 1054–1067 (2001)
4. S.L. Djajakesukma, B. Samali, H. Nguyen, Study of a semi-active stiffness damper under various earthquake inputs. Earthq. Eng. Struct. Dyn. **31**, 1757–1776 (2002)
5. G. Yan, L.L. Zhou, Integrated fuzzy logic and genetic algorithms for multi-objective control of structures using MR dampers. J. Sound Vib. **296**, 368–382 (2006)
6. F. Yi, S.J. Dyke, Structural control systems: performance assessment. Proc. Am. Control Conf. **1**, 14–18 (2000)
7. F.Y. Cheng, H. Jiang, K. Lou, *Smart Structures: Innovative Systems for Seismic Response Control* (CRC Press, Boca Raton, 2008)
8. Z. Liang, G.C. Lee, G.F. Dargush, J. Song, *Structural Damping: Applications in Seismic Response Modification* (CRC Press, Boca Raton, 2011)

9. N.R. Fisco, H. Adeli, Smart structures: part I—active and semi-active control. Sci. Iran. **18**, 275–284 (2011)
10. N.R. Fisco, H. Adeli, Smart structures: part II—hybrid control systems and control strategies. Sci. Iran. **18**, 285–295 (2011)
11. T.K. Datta, A state-of-the-art review on active control of structures. ISET J. Earthq. Technol. **40**, 1–17 (2003)
12. S. Korkmaz, A review of active structural control: challenges for engineering informatics. Comput. Struct. **89**, 2113–2132 (2011)
13. A.C. Nerves, R. Krishnan, Active control strategies for tall civil structures. Proc. IEEE Int. Conf. Ind. Electron. Control Instrum. **2**, 962–967 (1995)
14. T.T. Soong, S.F. Masri, G.W. Housner, An overview of active structural control under seismic loads. Earthq. Spectra **7**, 483–505 (1991)
15. J.N. Yang, T.T. Soong, Recent advances in active control of civil engineering structures. Probab. Eng. Mech. **3**, 179–188 (1988)
16. B.F. Spencer, S. Nagarajaiah, State of the Art of structural control. J. Struct. Eng. **129**, 845–856 (2003)
17. T.T. Soong, B.F. Spencer, Supplemental energy dissipation: state-of-the-art and state-of-the-practice. Eng. Struct. **24**, 243–259 (2002)
18. B.F. Spencer, M.K. Sain, Controlling buildings: a new frontier in feedback. IEEE Control Syst. Mag. Emerg. Technol. **17**, 19-35 (1997)
19. M.D. Symans, M.C. Constantinou, Semi-active control systems for seismic protection of structures: a state-of-the-art review. Eng. Struct. **21**, 469–487 (1999)
20. G. Kerschen, K. Worden, A.F. Vakakis, J.C. Golinva, Past, present and future of nonlinear system identification in structural dynamics. Mech. Syst. Signal Process. **20**, 505–592 (2006)
21. S.B. Kim, C.B. Yun, Sliding mode fuzzy control: theory and verification on a benchmark structure. Earthq. Eng. Struct. Dyn. **29**, 1587–1608 (2000)
22. Y. Tang, Active control of SDF systems using artificial neural networks. Comput. Struct. **60**, 695–703 (1996)
23. S. Thenozhi, W. Yu, Advances in modeling and vibration control of building structures. Annu. Rev. Control **37**(2), 346–364 (2013)
24. B. Jiang, X. Wei, Y. Guo, Linear quadratic optimal control in active control of structural vibration systems, in *Control and Decision Conference, 2010 Chinese*, vol 98, pp. 3546–3551 (2010)
25. Q.S. Li, D.K. Liu, J.Q. Fang, C.M. Tam, Multi-level optimal design of buildings with active control under winds using genetic algorithms. J. Wind Eng. Ind. Aerodyn. **86**, 65–86 (2000)
26. A.K. Chopra, *Dynamics of Structures: Theory and application to Earthquake engineering*, 2nd edn. (Prentice Hall, 2001)
27. Y.K. Wen, Method for random vibration of hysteretic systems. J. Eng. Mech. **102**, 249–263 (1976)
28. F. Ikhouane, V. Mañosa, J. Rodellar, Dynamic properties of the hysteretic Bouc-Wen model. Syst. Control Lett. **56**, 197–205 (2007)
29. F. Amini, M.R. Tavassoli, Optimal structural active control force, number and placement of controllers. Eng. Struct. **27**, 1306–1316 (2005)
30. M. Guney, E. Eskinat, Optimal actuator and sensor placement in flexible structures using closed-loop criteria. J. Sound Vib. **312**, 210–233 (2008)
31. W. Gawronski, Actuator and Sensor placement for structural testing and control. J. Sound Vib. **208**, 101–109 (1997)
32. W. Liu, Z. Hou, M.A. Demetriou, A computational scheme for the optimal sensor/actuator placement of flexible structures using spatial H_2 measures. Mech. Syst. Sig. Process. **20**, 881–895 (2006)
33. D.K. Liu, Y.L. Yang, Q.S. Li, Optimum positioning of actuators in tall buildings using genetic algorithm. Comput. Struct. **81**, 2823–2827 (2003)
34. L. Ljung, *System Identification Theory for the Users* (Prentice-Hall Inc, New Jersey, 1987)

35. H. Imai, C.B. Yun, O. Maruyama, M. Shinozuka, Fundamentals of system identification in structural dynamics. Probab. Eng. Mech. **4**, 162–173 (1989)
36. J. Zhang, P.N. Roschke, Active control of a tall structure excited by wind. J. Wind Eng. Ind. Aerodyn. **83**, 209–223 (1999)
37. D.P. Mandic, J.A. Chambers, *Recurrent Neural Networks for Prediction: Learning Algorithms, Architectures and Stability* (Wiley, Hoboken, 2001)
38. S.L. Hung, C.S. Huang, C.M. Wen, Y.C. Hsu, Nonparametric identification of a building structure from experimental data using wavelet neural network. Comput. Aided Civ. Infrastruct. Eng. **18**, 356–368 (2003)
39. F. Plestana, Y. Shtessel, V. Bregeaulta, A. Poznyak, New methodologies for adaptive sliding mode control. Int. J. Control **83**, 1907–1919 (2010)
40. B. Xu, Z. Wu, G. Chen, K. Yokoyama, Direct identification of structural parameters from dynamic responses with neural networks. Eng. Appl. Artif. Intell. **17**, 931–943 (2004)
41. M.J. Perry, C.G. Koh, Y.S. Choo, Modified genetic algorithm strategy for structural identification. Comput. Struct. **84**, 529–540 (2006)
42. W.H. Zhu, Velocity estimation by using position and acceleration sensors. IEEE Trans. Ind. Electron. **54**, 2706–2715 (2007)
43. D.M. Boore, Analog-to-Digital conversion as a source of drifts in displacements derived from digital recordings of ground acceleration. Bull. Seismol. Soc. Am. **93**, 2017–2024 (2003)
44. R. Adhikari, H. Yamaguchi, T. Yamazaki, Modal space sliding-mode control of structures. Earthq. Eng. Struct. Dyn. **27**, 1303–1314 (1998)
45. C.M. Casado, I.M. Díaz, J.D. Sebastián, A.V. Poncela, A. Lorenzana, Implementation of passive and active vibration control on an in-service footbridge. Struct. Control Health Monit. **20**, 70–87 (2013)
46. F.L. Lewis, D.M. Dawson, C.T. Abdallah, *Robot Manipulator Control: Theory and Practice*, 2nd edn. (Marcel Dekker, Inc, 2004)
47. S.F. Ali, A. Ramaswamy, Optimal fuzzy logic control for MDOF structural systems using evolutionary algorithms. Eng. Appl. Artif. Intell. **22**, 407–419 (2009)
48. K.J. Åström, T. Hagglund, Revisiting the Ziegler-Nichols step response method for PID control. J. Process Control **14**, 635–650 (2004)
49. Z.Q. Gu, S.O. Oyadiji, Application of MR damper in structural control using ANFIS method. Comput. Struct. **86**, 427–436 (2008)
50. Y.K. Thong, M.S. Woolfson, J.A. Crowe, B.R.H. Gill, D.A. Jones, Numerical double integration of acceleration measurements in noise. Measurement **36**, 73–92 (2004)
51. T.S. Edwards, Effects of aliasing on numerical integration. Mech. Syst. Sign. Process. **21**, 165–176 (2007)
52. K. Worden, Data processing and experiment design for the restoring force surface method, part I: integration and differentiation of measured time data. Mech. Syst. Sign. Process. **4**, 295–319 (1990)
53. H.P. Gavin, R. Morales, K. Reilly, Drift-free integrators. Rev. Sci. Instrum. **69**, 2171–2175 (1998)
54. J.G.T. Ribeiro, J.T.P. de Castro, J.L.F. Freire, Using the FFT-DDI method to measure displacements with piezoelectric, resistive and ICP accelerometers. Conf. Expo. Struct. Dyn. (2003)
55. J. Yang, J.B. Li, G. Lin, A simple approach to integration of acceleration data for dynamic soil-structure interaction analysis. Soil Dyn. Earthq. Eng. **26**, 725–734 (2006)
56. K.S. Park, H.M. Koh, C.W. Seo, Independent modal space fuzzy control of earthquake-excited structures. Eng. Struct. **26**, 279–289 (2004)
57. S.H. Razavi, A. Abolmaali, M. Ghassemieh, A weighted residual parabolic acceleration time integration method for problems in structural dynamics. J. Comput. Methods Appl. Math. **7**, 227–238 (2007)
58. R. Kelly, A tuning procedure for stable PID control of robot manipulators. Robotica **13**, 141–148 (1995)

59. C.M. Chang, B.F. Spencer, Active base isolation of buildings subjected to seismic excitations. Earthq. Eng. Struct. Dyn. **39**, 1493–1512 (2010)
60. M.D. Iuliis, C. Faella, Effectiveness analysis of a semiactive base isolation strategy using information from an early-warning network. Eng. Struct. **52**, 518–535 (2013)
61. K.C.S. Kwok, B. Samali, Performance of tuned mass dampers under wind loads. Eng. Struct. **17**, 655–667 (1995)
62. R. Saragih, Designing Active vibration control with minimum order for flexible structures. IEEE Int. Conf. Control Autom. 450–453 (2010)
63. J.T. Kim, H.J. Jung, I.W. Lee, Optimal structural control using neural networks. J. Eng. Mech. **126**, 201–205 (2000)
64. G. Miwada, O. Yoshida, R. Ishikawa, M. Nakamura, Observation records of base-isolated buildings in strong motion area during the 2011 off the Pacific Coast of Tohoku earthquake, in *Proceedings of the International Symposium on Engineering Lessons Learned from the 2011 Great East Japan Earthquake* (2012), pp. 1017–1024
65. D.H. Kim, Neuro-control of fixed offshore structures under earthquake. Eng. Struct. **31**, 517–522 (2009)
66. T.T. Soong, *Active Structural Control: Theory and Practice* (Longman, New York, 1990)
67. T.T. Soong, A.M. Reinhorn, Y.P. Wang, R.C. Lin, Full-scale implementation of active control-I: design and simulation. J. Struct. Eng. **117**, 3516–3536 (1991)
68. J.C.H. Chang, T.T. Soong, Structural control using active tuned mass damper. J. Eng. Mech. ASCE **106**, 1091–1098 (1980)
69. A. Alavinasab, H. Moharrami, Active control of structures using energy-based LQR method. Comput. Aided Civ. Infrastruct. Eng. **21**, 605–611 (2006)
70. A. Forrai, S. Hashimoto, H. Funato, K. Kamiyama, Structural control technology: system identification and control of flexible structures. Comput. Control Eng. J. **402**, 1–40 (2001)
71. D. Hrovat, P. Barak, M. Rabins, Semi-active versus passive or active tuned mass dampers for structural control. J. Eng. Mech. **109**, 691–705 (1983)
72. Z. Xu, A.K. Agrawal, J.N. Yang, Semi-active and passive control of the phase I linear base-isolated benchmark building model. Struct. Control Health Monit. **13**, 626–648 (2006)
73. J. Pandya, Z. Akbay, M. Uras, H. Aktan, *Exp. Implement. Hybrid Control* (Proceedings of Structures Congress XIV, Chicago, 1996)
74. B.F. Spencer, S.J. Dyke, M.K. Sain, J.D. Carlson, Phenomenological model of a magnetorheological damper. J. Eng. Mech. ASCE **123**, 230–238 (1997)
75. Y.L. Xu, B. Chen, Integrated vibration control and health monitoring of building structures using semi-active friction dampers: Part I-methodology. Eng. Struct. **30**, 1789–1801 (2008)
76. W. Park, K.S. Park, H.M. Koh, Active control of large structures using a bilinear pole-shifting transform with H_∞ control method. Eng. Struct. **30**, 3336–3344 (2008)
77. O.I. Obe, Optimal actuators placements for the active control of flexible structures. J. Math. Anal. Appl. **105**, 12–25 (1985)
78. R. Guclu, H. Yazici, Vibration control of a structure with ATMD against earthquake using fuzzy logic controllers. J. Sound Vib. **318**, 36–49 (2008)
79. R. Guclu, Sliding mode and PID control of a structural system against earthquake. Math. Comput. Model. **44**, 210–217 (2006)
80. T.L. Teng, C.P. Peng, C. Chuang, A study on the application of fuzzy theory to structural active control. Comput. Methods Appl. Mech. Eng. **189**, 439–448 (2000)
81. V.I. Utkin, *Sliding Modes in Control and Optimization* (Springer, Berlin, 1990)
82. C.W. Chen, Modeling and control for nonlinear structural systems via a NN-based approach. Expert Syst. Appl. **36**, 4765–4772 (2009)
83. H. Du, N. Zhang, H_∞ control for buildings with time delay in control via linear matrix inequalities and genetic algorithms. Eng. Struct. **30**, 81–92 (2008)
84. K. Seto, A structural control method of the vibration of flexible buildings in response to large earthquake and strong winds, in *Proceedings of the 35th Conference on Decision and Control* (Kobe, Japan, 1996)

85. S. Pourzeynali, H.H. Lavasani, A.H. Modarayi, Active control of high rise building structures using fuzzy logic and genetic algorithms. Eng. Struct. **29**, 346–357 (2007)
86. Y.K. Wen, J. Ghaboussi, P. Venini, K. Nikzad, Control of structures using neural networks. Smart Mater. Struct. **4**, 149–157 (1995)
87. H.C. Cho, M.S. Fadali, M.S. Saiidi, K.S. Lee, Neural network active control of structures with earthquake excitation. Int. J. Control Autom. Syst. **2**, 202–210 (2005)
88. D.H. Kim, D. Kimb, S. Chang, H.Y. Jung, Active control strategy of structures based on lattice type probabilistic neural network. Probab. Eng. Mech. **23**, 45–50 (2008)
89. A. Madan, Vibration control of building structures using self-organizing and self-learning neural networks. J. Sound Vib. **287**, 759–784 (2005)
90. J. Liu, K. Xia, C. Zhu, Structural vibration intelligent control based on magnetorheological damper, in *International Conference on Computational Intelligence and Software Engineering* (2009), pp. 1–4
91. A. Tani, H. Kawamura, S. Ryu, Intelligent fuzzy optimal control of building structures. Eng. Struct. **20**, 184–192 (1998)
92. E. Reithmeier, G. Leitmann, Structural vibration control. J. Franklin Inst. **338**, 203–223 (2001)
93. M. Aldawod, F. Naghdy, B. Samali, K.C.S. Kwok, Active control of wind excited structures using fuzzy logic, in *IEEE International Fuzzy Systems Conference Proceedings* (1999), pp. 72–77
94. K. Yeh, W.L. Chiang, D.S. Juang, Application of fuzzy control theory in active control of structures, in *IEEE Proceeding NAFIPS/IFIS/NASA* (1994), pp. 243–247
95. S.K. Yalla, A. Kareem, J.C. Kantor, Semi-active tuned liquid column dampers for vibration control of structures. Eng. Struct. **23**, 1469–1479 (2001)
96. K.S. Park, H.M. Koh, S.Y. Ok, Active control of earthquake excited structures using fuzzy supervisory technique. Adv. Eng. Softw. **33**, 761–768 (2002)
97. K.M. Choi, S.W. Cho, D.O. Kim, I.W. Lee, Active control for seismic response reduction using modal-fuzzy approach. Int. J. Solids Struct. **42**, 4779–4794 (2005)
98. D. Das, T.K. Datta, A. Madan, Semiactive fuzzy control of the seismic response of building frames with MR dampers. Earthq. Eng. Struct. Dyn. **41**, 99–118 (2012)
99. N.D. Duc, N.L. Vu, D.T. Tran, H.L. Bui, A study on the application of hedge algebras to active fuzzy control of a seism-excited structure. J. Vib. Control, 1–15 (2011)
100. P.J. Fleming, R.C. Purshouse, Evolutionary algorithms in control systems engineering: a survey. Control Eng. Pract. **10**, 1223–1241 (2002)
101. K.S. Park, H.M. Koh, Preference-based optimum design of an integrated structural control system using genetic algorithms. Adv. Eng. Softw. **35**, 85–94 (2004)
102. L.J. Li, MRF absorber damping control for building structural vibration response by means of genetic optimum algorithm. Adv. Mater. Res. **219–220**, 1133–1137 (2011)
103. B. Zhao, X. Lu, M. Wu, Z. Mei, Sliding mode control of buildings with base-isolation hybrid protective system. Earthq. Eng. Struct. Dyn. **29**, 315–326 (2000)
104. M. Allen, F.B. Zazzera, R. Scattolini, Sliding mode control of a large flexible space structure. Control Eng. Pract. **8**, 861–871 (2000)
105. S.M. Nezhad, F.R. Rofooei, Decentralized sliding mode control of multistory buildings. Struct. Des. Tall Spec. Build. **16**, 181–204 (2007)
106. O. Yakut, H. Alli, Neural based sliding-mode control with moving sliding surface for the seismic isolation of structures. J. Vib. Control **17**, 2103–2116 (2011)
107. Z. Li, Z. Deng, Z. Gu, New sliding mode control of building structure using RBF neural networks. Chin. Control Decis. Conf. 2820–2825 (2010)
108. H. Alli, O. Yakut, Fuzzy sliding-mode control of structures. Eng. Struct. **27**, 277–284 (2005)
109. A.P. Wang, Y.H. Lin, Vibration control of a tall building subjected to earthquake excitation. J. Sound Vib. **299**, 757–773 (2007)

Chapter 3
Position and Velocity Estimation

Abstract In this chapter, offset cancellation and high-pass filtering techniques are combined effectively to solve common problems in numerical integration of acceleration signals in real-time applications. The integration accuracy is improved compared with other numerical integrators.

Keywords Numerical integration · Position estimation of vibration signals

In structural vibration control, relative positions and velocities are not easily measured directly because they require fixed reference positions in a building, which is difficult during seismic events. On the other hand, it is very easy to obtain acceleration signal from an accelerometer.

Accelerometers are very popular sensors in machinery and building health monitoring, structural vibration control, transport, and even personal electronic devices, because the structure of the accelerometer is very simple and it does not need any relative reference point. In fact, most of acceleration and tilt measurements use accelerometers. A comparison study on the performance of position, velocity, and acceleration sensors can be found in [1–3]. In order to design a PID or a state-space-based control via accelerometers, velocity and position estimations are needed. Observers like Kalman filters are popular methods to estimate the velocity and position from an acceleration signal [4, 5]. These approaches work well when the measurements are corrupted by Gaussian noise. However, they need prior knowledge of the plant's parameters, for instance the mass, damping, and stiffness of the building structure.

Integration is the most direct method to obtain the velocity and position from acceleration. Since the measured acceleration signal from an accelerometer contains offset and low-frequency noise, it is not appropriate to integrate the acceleration signal directly. Numerical integrators can be designed in the time domain [6–8] and in the frequency domain [9]. However, it is difficult to use frequency domain techniques for online integration [10].

The drift is the major problem in numerical integration, which is caused by unknown initial condition and the offset of the accelerometer. An ideal integrator amplifies direct current (DC) signals. To avoid drift during the integration it must be

© The Author(s) 2016

W. Yu and S. Thenozhi, *Active Structural Control with Stable*
Fuzzy PID Techniques, SpringerBriefs in Applied Sciences and Technology,
DOI 10.1007/978-3-319-28025-7_3

removed. Since the behavior of a first-order low-pass filter is similar to the behavior of an ideal integrator, the former may replace the latter in some instance. In order to remove the low-frequency offsets, a high-pass filter can be used [6]. The drift can also be canceled out by a feedback method. An advantage of their integrator is that its estimation error converges to zero exponentially. However, the main problem of these numerical integrators is that they have to use large time constants to avoid the drift. The behavior of a filter with a large time constant is far from an ideal integrator, which will result a reduced integration accuracy. In [10], the drift caused by unknown initial conditions is eliminated via a frequency domain transformation. A Butterworth-type filter is used as a numerical integrator in [11].

The baseline correction is an alternative method to avoid drift during integration. In [12], a polynomial baseline correction was applied to cancel out the offset. The polynomial curve is fitted using the least square method. The low-frequency components in the accelerometer were estimated and removed offline. In [7], a calibration technique and an initial velocity estimation were used to remove the integration error in double integrators. In [8], a weighted residual parabolic integration is proposed, where the position is assumed to be a fourth-order polynomial function of the acceleration. The main problem of these baseline correction integrators is that the low-frequency noise has to be removed by a special window filter, which has to be designed only for a particular input acceleration signal. Due to this reason it cannot be useful for online estimation, where the input signal frequency is unknown.

This chapter addresses all the critical sources of offsets and noise that can cause drift during the integration and a mathematical model for the accelerometer output signal is provided. An offset cancellation filter (OCF) is proposed, which removes the DC components present in the accelerometer output. In order to avoid the drift caused by low-frequency noise signals, a special high-pass filter is designed. A frequency domain method is used to estimate the low-frequency noise components present in the accelerometer output. The high-pass filter is designed offline according to these noise components. Since the OCF reduces the number of high-pass filtering stages, the resulting phase error has been reduced. The proposed numerical integrator combines the OCF and a high-pass filter. It is successfully applied on a linear servo actuator and on a shake table. The real-time experimental results validate the proposed method.

3.1   Numerical Integrator for Accelerometers

The accelerometer can be regarded as a single-degree-of-freedom (SDOF) mechanical system [13]. It is often modeled by a simple mass m, called proof mass, attached to a spring of stiffness k, and a dashpot with damping coefficient c, see Fig. 3.1. The inertial force acting on the proof mass is given by

$$F = m(\ddot{x}(t) + \ddot{x}_m(t)) \qquad (3.1)$$

Fig. 3.1 Mechanical model
of an accelerometer

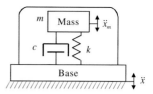

where $\ddot{x}(t)$ is the acceleration acting on the accelerometer and $\ddot{x}_m(t)$ is the relative
acceleration of the proof mass with respect to the base.

The dynamics of an accelerometer subjected to an acceleration $\ddot{x}(t)$ using New-
ton's second law is given by

$$m\ddot{x}_m(t) + c\dot{x}_m(t) + kx_m(t) = -m\ddot{x}(t) \qquad (3.2)$$

The deflection due to the acceleration is sensed and converted into an equivalent
electrical signal. This conversion can be represented using a constant gain termed as
the accelerometer gain, k_a,

$$\ddot{x}_m(t) + 2\zeta\omega_n\dot{x}_m(t) + \omega_n^2 x_m(t) = k_a\ddot{x}(t) \qquad (3.3)$$

However, the accelerometer measures the input acceleration with a slight change in
its amplitude defined by its gain k_a and phase, which are normally negligible [14].

Other than the input acceleration, the accelerometer output signal $a(t)$ contains
offset and noise. An accelerometer has a bias, named $0g$-offset, which is measured
under the absence of motion or gravity ($0g$). The $0g$-offset is normally equal to the
half of its power supply ($V_{dd}/2$). This offset may vary from one sensor to another. The
main causes of this variation are the sensing material, temperature, supply voltage
deviation, mechanical stress, and trim errors [15]. The knowledge of this offset error
can be useful in removing the bias from the acceleration signal effectively.

The accelerometer output signal can be represented as [13, 16]

$$a(t) = k_a\ddot{x}(t) + w(t) + \varphi \qquad (3.4)$$

where $w(t)$ is the noise and disturbance effects on the measurement, and φ denotes
the $0g$-offset.

Mathematically, the velocity $\dot{x}(t)$ and position $x(t)$ are calculated by integrating
the acceleration $\ddot{x}(t)$

$$
\begin{aligned}
\dot{x}(t) &= \int_0^t \ddot{x}(\tau)d\tau + \dot{x}(0) \\
x(t) &= \int_0^t \int_0^\tau \ddot{x}(\tau)d\tau dt + \dot{x}(0)t + x(0)
\end{aligned}
\qquad (3.5)
$$

where $\dot{x}(0)$ and $x(0)$ are the initial velocity and position, respectively.

In discrete time, the numerical integration is performed to get an approximation by applying the numerical interpolation,

$$\int_{t_0}^{t_n} \ddot{x}(t)dt \approx \sum_{i=1}^{n} \left[\frac{\ddot{x}(i-1) + \ddot{x}(i)}{2} \right] \Delta t \qquad (3.6)$$

There exist several types of numerical integration techniques in the time domain and in the frequency domain. Trapezium rule, Simpson's rule, Tick's rule [9], and rectangular rule [7] are popular time domain integration techniques. Fourier transformation is a frequency domain method, which is a better tool for dealing with nonperiodic functions. In the frequency domain, the Fourier Transform $\mathcal{F}$ of acceleration $H_{\ddot{x}}(\omega)$ is [9]

$$H_{\ddot{x}}(\omega) = \mathcal{F}\{\ddot{x}(t)\} = \int_{-\infty}^{\infty} \ddot{x}(t)e^{-i\omega t} dt \qquad (3.7)$$

The velocity and position are obtained by dividing $H_{\ddot{x}}(\omega)$ by $i\omega$ and $(i\omega)^2$, respectively. These are then converted back to the time domain by using the inverse Fourier Transform. However, this method does not have a good low-frequency characteristics, for example, the leakage problem.

3.2 Novel Numerical Integrator

Time integration of the acceleration is a straightforward solution for estimating the velocity and position. However, there are four problems affecting the performance of numerical integrators:

(1) *Bias*: The behavior of a numerical integrator is similar to a low-pass filter. It amplifies the low-frequency components, reduces the magnitude of high-frequency signals, and causes a phase error. Therefore, any bias in the acceleration measurement results in integration drift. Ambient temperature change is another major offset source. A low resolution analog-to-digital converter (ADC) also adds an offset when the acceleration is slow compared with the quantization level of the analog-to-digital conversion [17].

(2) *Noise*: There are different sources of noise in the accelerometer, which is generally modeled as white noise. Integrating these noisy signals result a large error in the velocity and position estimations. The root mean square (RMS) of the position estimation error $e_x(t)$ of the acceleration signal with a bias $\bar{\varphi}$ can be approximated as [7]

$$\text{RMS}\{e_x(t)\} = \frac{1}{2}\bar{\varphi}t^2 \qquad (3.8)$$

which grows at the rate of t^2, where t is the time.

(3) *Aliasing*: It is caused when digitizing an analog signal with a constant sampling frequency, because the frequency components above the Nyquist rate in ADC are folded back into the frequency of interest. When ADC produces aliasing, the output signal in (3.4) can be represented as follows:

$$a(t) = k_a \ddot{x}(t) + \ddot{x}_s(t) + w(t) + \varphi \tag{3.9}$$

where $\ddot{x}_s(t)$ is the aliasing component due to sampling. Aliasing phenomena can cause low-frequency errors and is amplified during the integration process [18]. This error can be minimized using an anti-aliasing filter between the accelerometer and data acquisition unit. Another method is to use a large and constant sampling rate.

(4) *Integration technique*: The numerical integration methods like the Trapezium rule, Simpson's rule, and Tick's rule do not have good properties at low frequencies. It is also shown that the Simpson's and Tick's rules are unstable at high frequencies [9].

The motivation for the present work lies in these considerations and we propose a strategy to overcome all the problems mentioned above. To achieve this, different signal processing techniques have been applied. Among the four problems mentioned above, the first one (due to the presence of bias) is more critical. We propose a new offset cancellation filter, which can remove the offset from the acceleration output. This filter is basically a numerical algorithm, which will remove the constant and slowly varying signals from the measured. The design procedure of this filter is discussed in the Sect. 3.3.1.

The most straightforward way to remove noise is to use a filter. Here, we use a low-pass filter at the output of the accelerometer to remove any high-frequency noise signals. A Sallen–Key high-pass filter is proposed that accomplishes two jobs: (1) to attenuate the low-frequency noise, (2) to avoid drifting at the integration output by providing damping. Here the filter order and cutoff frequency is selected in such a way that the frequency of interest has minor effects by the filtering. This filter design is explained in the Sect. 3.3.2.

The useful frequency of the building structure lies in the low-frequency spectrum. Based on the observations from [9], the methods like Trapezium rule, Simpson's rule, and Tick's rule should not be used for integrating low-frequency signals like the structural acceleration. Dormand–Prince is a popular method, where the solution is computed using a higher order formula, which results in an accurate and stable result. Here we use this scheme for performing the integration operation.

The final problem is concerned with the data acquisition hardware. An anti-aliasing filter is used between the accelerometer and data acquisition unit for minimizing the aliasing effect. Moreover, we use a high-resolution ADC for the analog-to-digital conversion and the sampling is done at a high rate using a dedicated clock source. More details on hardware components are provided in experiments section.

3.2.1 Offset Cancellation Filter

The offset voltage present in the accelerometer is one of the main causes of integration drift. Figure 3.2 shows drifting of the integrated signal in the presence of offset. Here a constant bias of 0.05 m/s^2 is added to a 1 m/s^2 acceleration signal and integrated two times. Compared to the first integrated signal the second integrated signal drifts very fast. It evidently indicates that, if there exists any bias in the acceleration signal the resulting position estimation using double integration will drift rapidly with time.

In [12], polynomial baselines have been used for removing the offset present in the estimations, which is given by

$$\begin{aligned}
p(t) &= a_1 t^4 + a_2 t^3 + a_3 t^2 + a_4 t \\
\dot{p}(t) &= 4a_1 t^3 + 3a_2 t^2 + 2a_3 t \\
\ddot{p}(t) &= 12a_1 t^2 + 6a_2 t + 2a_3
\end{aligned} \tag{3.10}$$

where $p(t)$, $\dot{p}(t)$, and $\ddot{p}(t)$ are the baselines for position, velocity, and acceleration, respectively. Once the above baselines are determined, they are then subtracted from the estimations to remove the offset. The acceleration data is measured first and the corresponding polynomial coefficients $(a_1 - a_4)$ are calculated offline using the least square curve fitting method. For that reason, this method cannot be applied for online estimation. On the other hand, this scheme gives a good estimation for the offline data.

The offset cancellation is also called offset calibration, where the $0g$-offset voltage under $0g$-motion is removed from the accelerometer output [15]. Many practical applications use an electronic voltage compensator for removing this offset. Since this offset changes with time, this circuit needs frequent calibration, which may be difficult in some applications. Another option is to use a high-pass filter in order to remove the low-frequency DC components. The major drawback of this approach is that the high-pass filter introduces a phase error in the cutoff frequency range.

The proposed approach uses a numerical filter to remove the offset. The advantage of this method is that it does not add any phase error to the output. The ideal integrator generates unbounded output signal, if its input signal contains pure DC components.

Fig. 3.2 Integration in the presence of offset

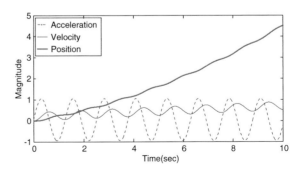

Since the OCF is able to remove the DC components completely, an ideal integrator can be used for the integration operation.

The initial condition for the acceleration is assumed to be zero; $\ddot{x}(0) = 0$ so that

$$a(0) = \varphi \tag{3.11}$$

If $a(t)$ in (3.9) is delayed for one sample time (δ_d), it becomes

$$a(t - \delta_d) = k_a \ddot{x}(t - \delta_d) + \ddot{x}_s(t - \delta_d) + w(t - \delta_d) + \varphi \tag{3.12}$$

Consider a variable $\epsilon(t)$

$$\epsilon(t) = a(t) - a(t - \delta_d); \quad \epsilon(0) = 0 \tag{3.13}$$

Now the offset-free acceleration $\ddot{\hat{x}}(t)$ can be found as

$$\ddot{\hat{x}}(t) = \sum_{i=0}^{n} \epsilon(t - i); \quad \ddot{\hat{x}}(0) = 0 \tag{3.14}$$

$$= k_a \ddot{x}(t) + \ddot{x}_s(t) + w(t) \tag{3.15}$$

It is clear that the above algorithm can remove the pure DC component (φ) completely. Here, it is assumed that the acceleration signal is unknown but bounded, i.e.,

$$|a(t)| \leq \bar{a} \ \forall t \geq 0$$

where $\bar{a}$ is a finite, positive constant. Therefore, the $\ddot{\hat{x}}(t)$ is also bounded.

If the offset changes slowly with time, then it is represented as $\varphi(t)$. Since the offset frequency is close to 0 Hz, we use the following two ways to reduce the effect of $\varphi(t)$ in the estimation.

(1) The scheme is to identify the slowly changing signal close to 0 Hz and to remove it from the acceleration signal. If the offset is changing slowly, the resulting rate of change of acceleration signal $a(t)$, from one sample data to the next sample will be very small, i.e., small $\epsilon(t)$. This small $\epsilon(t)$ is identified and removed from the acceleration signal to nullify the slowly changing $\varphi(t)$ as shown below.

$$\epsilon(t) = \begin{array}{ll} 0 & \text{if} \ \ \epsilon(t) < \epsilon_{\min} \\ \epsilon(t) & \text{if} \ \ \epsilon(t) \geq \epsilon_{\min} \end{array} \tag{3.16}$$

where $\epsilon_{\min}$ is the smallest value of $\epsilon(t)$ to be removed. Due to the above scheme, we can write $\left|\ddot{\hat{x}}(t)\right| \leq |a(t)|$, which shows that boundness is still preserved. For example, consider an offset changing very slowly at a constant rate of v with time, so that $\varphi(t) = \varphi + vt$. In this case, the offset can be removed by choosing $\epsilon_{\min} \geq v$.

The change in the offset is due to different causes like the temperature changes. As the rate of change of offset is different from one accelerometer to the another, the ϵ_{min} will also differ.

(2) The OCF reset is performed such that the slowly varying offset is canceled out more effectively. The reset should be done in the absence of motion. Once the reset is carried out the initial acceleration $a(0)$ corresponds to the new offset, which will be removed by the OCF. Simply speaking, the OCF removes very slowly changing signals from the acceleration signal.

In [6], a low-pass filter and a high-pass filter is used for removing the effect of the offset. Instead of using an ideal integrator a low-pass filter is used as follows:

$$L(s) = \frac{1}{s + \tau^{-1}} \tag{3.17}$$

By increasing the filter time constant τ, the output offset can be reduced, but doing that will add phase error to the signal. Here the offset is removed using the proposed OCF and then the ideal integrator can be used without making it unstable. Moreover, the OCF reduces the offset without adding any phase error.

In practice, the term $a(0) \neq 0$, which can be represented as

$$a(0) = \varphi + \vartheta \tag{3.18}$$

where $\vartheta \ll \varphi$ is from the noise $w(t)$ and other noise sources. Then, the output of the OCF is

$$\ddot{\hat{x}}(t) = k_a \ddot{x}(t) + \ddot{x}_s(t) + w(t) + \vartheta \tag{3.19}$$

The term ϑ is removed using a second-order high-pass filter as discussed in the section below.

3.2.2   High-Pass Filtering for Drift Attenuation

The OCF removes the DC components efficiently. However, it cannot deal with other low-frequency noises, which also cause drift in the integrator. To remove the low-frequency components in (3.9), we use a second-order high-pass filter. The transfer function of a second-order unity-gain Sallen–Key high-pass filter is

$$G(s) = \frac{s^2}{s^2 + 2\tau^{-1}s + \tau^{-2}} \tag{3.20}$$

where τ is estimated using the Fast Fourier Transform (FFT). The cutoff frequency (f_c) of the filter is

$$f_c = \frac{1}{2\pi\tau} \tag{3.21}$$

The FFT gives the frequency distribution of the accelerometer output signal under $0g$-motion. The cutoff frequency of the filter (3.20) is calculated based on the noise distribution. The high-pass filter design is performed offline. During design, the effect of the filter on the low-frequency information should be considered. The cutoff frequency should be selected in such a way that it would not attenuate the low-frequency information data. It will be a good practice to use low-noise accelerometers, so that the filter cutoff frequency can be kept low. Once the filter is designed it can deal with the acceleration signals above its cutoff frequency, so that a wide range of building structure frequencies, which makes them capable of performing online estimation.

The scheme of the proposed numerical integrator is shown in Fig. 3.3. Initially, the high-frequency noise present in the accelerometer output signal is attenuated using a low-pass filter. This filtered acceleration signal is passed through the OCF for removing the offset as explained in (3.12)–(3.16). This signal is integrated to get velocity estimation and then given to a high-pass filter for removing the low-frequency noise. The integrator and high-pass filter are cascaded, which gives

$$G(s) = \frac{s}{s^2 + 2\tau^{-1}s + \tau^{-2}} \tag{3.22}$$

Then the velocity estimation $\dot{\hat{x}}(t)$ can be expressed as

$$\dot{\hat{x}}(t) = \mathcal{L}^{-1}\left[G(s)\left(\mathcal{L}\left[\ddot{\hat{x}}(t)\right]\right)\right] \tag{3.23}$$

where $\mathcal{L}$ is the Laplace transform operator. A zero initial condition is considered for both position and velocity, which is reasonable in the case of building structure in the absence of any excitation. Similarly, the position estimation $\hat{x}(t)$ is obtained as

$$\hat{x}(t) = \mathcal{L}^{-1}\left[G(s)\left(\mathcal{L}\left[\dot{\hat{x}}(t)\right]\right)\right] \tag{3.24}$$

Fig. 3.3 Scheme of the proposed numerical integrator

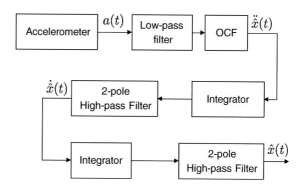

The anti-aliasing filter and oversampling technique is used to minimize the aliasing effects. Sometimes, the noise and information signal frequencies may be in the same band. In that case it will be difficult to remove these noise signals.

3.3 Experimental Results

A linear servo actuator mechanism and a shaking table are used here to evaluate the velocity and position estimations. The accelerometer is Summit Instruments 13203B. The $0g$-offset of the accelerometer is 2.44 V and the temperature drift is 3.2 mg/°C. The built-in temperature sensor in the accelerometer is utilized for compensating this temperature effect. The accelerometer output in $0g$-motion is integrated and the output drift is shown in Fig. 3.4.

ServoToGo Model II data acquisition card is employed to acquire the acceleration signal. The data acquisition card uses a 13-bit ADC. The acceleration signal is recorded at a sampling rate of 1 ms. In order to assure a constant sampling interval, a dedicated clock source is used for the data acquisition card. This will help in reducing the low-frequency noise in the acquired acceleration signal. The Dormand–Prince method is chosen for the integration.

The Fourier spectrum of the accelerometer $0g$-motion output signal is plotted using FFT, see Fig. 3.5. From the plot it is clear that the accelerometer has a measurement noise close to 0 Hz. The high-pass filter is designed ($f_c = 0.16$ Hz; $\tau = 1$) to attenuate these noise signals. As the natural frequency of the mechanical structure is 7.7 Hz, the above filter does not affect this frequency. A low-pass filter is used in the accelerometer output for attenuating the signals above 30 Hz. As the position sensor is available for both experiments, we use the measured position data to compare that with the position estimation obtained using the numerical integrator.

Fig. 3.4 Drift in the integration output

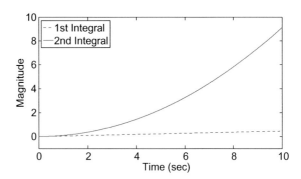

Fig. 3.5 Fourier spectra of
the acceleration signal for
zero motion

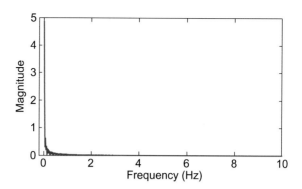

3.3.1 Linear Servo Actuator

Once the parameters of the proposed numerical integrator are calculated, the experiments were carried out to evaluate the velocity and position estimations. The linear servomechanism (STB1108, Copley Controls Corp.) is driven using a digital servo drive (Accelnet Micro Panel, Copley Controls Corp.). The servo tube comes with an integrated position sensor with a resolution of 8 μm, which is used here as the reference for verifying the estimated position. The servomechanism is actuated using basic sinusoidal signals and the corresponding acceleration is measured with the accelerometer. The accelerometer is mounted on the actuator, where its sensitive axis is placed parallel to the direction of actuator motion, see Fig. 3.6.

A 4 Hz sinusoidal signal, a signal composed with 6, 7, and 8 Hz and a signal composed with 2, 4, 6, and 8 Hz, is used here to excite the linear actuator. The acceleration of the actuator is measured using the accelerometer and fed to the OCF ($\epsilon_{min} = 0.001$) for removing the offset. Figure 3.7 shows the Fourier spectra of both the measured and filtered acceleration signals. We can see that the low-frequency

Fig. 3.6 Linear
servomechanism

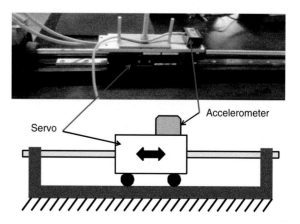

Fig. 3.7 Fourier spectra of
the acceleration signal before
and after filtering using OCF

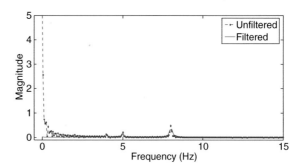

noise signals are removed. This filtered acceleration signal is then fed to the proposed
integrator for estimating the velocity and position. The position estimations for a 4 Hz
sine wave, signal composed with 6, 7, and 8 Hz, and signal composed with 4, 6, and
8 Hz, are shown in Figs. 3.8, 3.9, and 3.10, respectively.

The effect of the proposed numerical integrator on the input signal frequency
characteristics is studied by plotting its Fourier spectra. A sinusoidal signal composed
with 6, 7, and 8 Hz is used here to excite the linear actuator. The FFT diagram of
the measured and estimated position is generated, see Fig. 3.11. As one can see from
the figure that the frequency information is not affected, except in the low-frequency

Fig. 3.8 Comparison of the
measured and estimated
positions with 4Hz signal

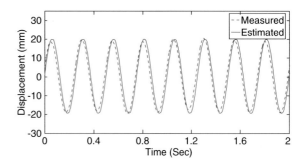

Fig. 3.9 Comparison of the
measured and estimated
positions with 6Hz signal

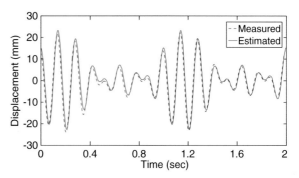

Fig. 3.10 Comparison of the measured and estimated positions with 8Hz signal

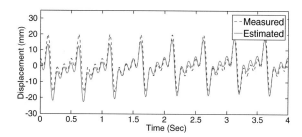

Fig. 3.11 Comparison of the measured and estimated position data using Fourier spectra

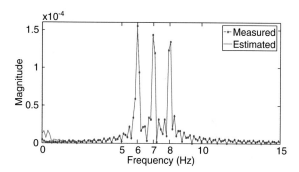

range. This low-frequency error is caused due to the presence of bias and noise in the accelerometer output.

3.3.2 Shaking Table

A shaking table prototype is used to verify the estimation during the earthquake excitation. The prototype is actuated using the earthquake signal and the structure acceleration is measured, which is then used to estimate the structure velocity and position. A linear magnetic encoder (LM15) position sensor with a resolution of 50 μm is used here for measuring the structure position. The mechanical structure base is connected to the electrohydraulic shaker (FEEDBACK EHS 160), which is used to generate the earthquake signals. The experimental setups are shown in Figs. 3.12 and 3.13.

The natural and forced responses of the mechanical structure are evaluated. The excitation signal is generated manually by knocking the structure with a hammer to bring out its natural response. The measured acceleration signal is fed to the proposed integrator and the position is estimated, which is shown in Fig. 3.14. In order to perform a comparison this figure also includes the estimation performed using the integrator proposed in [6] ($\alpha = 1$, $\beta = 0.2$, $K = 1$, and $\tau = 1$).

Finally, the October 17, 1989 Loma Prieta East-West earthquake signal is generated using the electrohydraulic shaker and the resulting acceleration on the

Fig. 3.12 Shaking table
experimental setup

Fig. 3.13 Schematic of the
shaking table setup

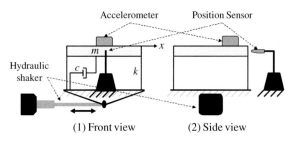

Fig. 3.14 Comparison of the
measured and estimated
positions with an artificial
signal

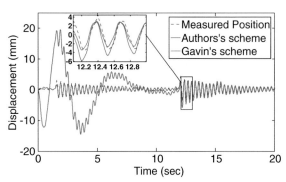

mechanical structure is measured. The corresponding position estimation is shown
in Fig. 3.15.

From the above experiments, it can be seen that the proposed integrator is able
to estimate the velocity and position with a reasonable level of accuracy. Still, there
exists some error between the estimated and measured position. This error is caused
due to the phase error, introduced by the high-pass filter, which resulted in a small
phase error. But it is found that the estimation obtained using the proposed integrator
is adequate for the structural control and health monitoring applications.

In this chapter, it is assumed that the building structure natural frequencies lie
between 1 and 20 Hz. As mentioned earlier, the high-pass filter introduces phase
errors in the cutoff frequency region. The resulting error is variable with the signal

Fig. 3.15 Comparison of the measured and estimated positions with Loma Prieta earthquake signal

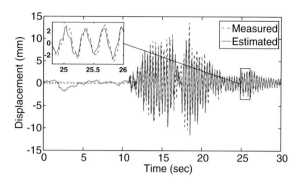

Fig. 3.16 Comparison of the measured and estimated position data obtained using different high-pass filters

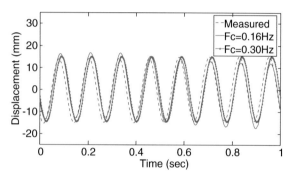

frequency. If the structure natural frequency is close to the high-pass filter cutoff frequency, then the estimation is affected due to the phase error introduced by the filter. The knowledge about structure natural frequency can be considered in the high-pass filter design. In Fig. 3.16 the position estimation of a 8 Hz sinusoidal signal obtained using two different high-pass filters is shown. The first filter has a cutoff frequency of 0.16 Hz and the second filter has 0.3 Hz. The estimation obtained using the first filter have some low-frequency noise. This problem is solved if the second filter is used for the estimation. Moreover, this filter cutoff frequency is far from the input signal frequency.

References

1. G.W. Housner, L.A. Bergman, T.K. Caughey, A.G. Chassiakos, R.O. Claus, S.F. Masri, R.E. Skelton, T.T. Soong, B.F. Spencer, J.T.P. Yao, Structural control: past, present and future. J. Eng. Mech. **123**, 897–974 (1997)
2. T. Balendra, C.M. Wang, N. Yan, Control of wind-excited towers by active tuned liquid column damper. Eng. Struct. **23**, 1054–1067 (2001)
3. Y.L. Xu, Parametric study of active mass dampers for wind-excited tall buildings. Eng. Struct. **18**, 64–76 (1996)
4. Z.Q. Gu, S.O. Oyadiji, Application of MR damper in structural control using ANFIS method. Comput. Struct. **86**, 427–436 (2008)

5. Y.L. Xu, B. Chen, Integrated vibration control and health monitoring of building structures using semi-active friction dampers: Part I-methodology. Eng. Struct. **30**, 1789–1801 (2008)
6. H.P. Gavin, R. Morales, K. Reilly, Drift-free integrators. Rev. Sci. Instrum. **69**, 2171–2175 (1998)
7. Y.K. Thong, M.S. Woolfson, J.A. Crowe, B.R.H. Gill, D.A. Jones, Numerical double integration of acceleration measurements in noise. Measurement **36**, 73–92 (2004)
8. S.H. Razavi, A. Abolmaali, M. Ghassemieh, A weighted residual parabolic acceleration time integration method for problems in structural dynamics. J. Comput. Methods Appl. Math. **7**, 227–238 (2007)
9. K. Worden, Data processing and experiment design for the restoring force surface method, Part I: integration and differentiation of measured time data. Mech. Syst. Signal Process. **4**, 295–319 (1990)
10. J.G.T. Ribeiro, J.T.P. de Castro, J.L.F. Freire, Using the FFT- DDI method to measure displacements with piezoelectric, resistive and ICP accelerometers, in *Conference and Exposition on Structural Dynamics* (2003)
11. S.F. Masri, L.H. Sheng, J.P. Caffrey, R.L. Nigbor, M. Wahbeh, A.M. Abdel-Ghaffar, Application of a web-enabled real-time structural health monitoring system for civil infrastructure systems. Smart Mater. Struct. **13**, 1269–1283 (2004)
12. J. Yang, J.B. Li, G. Lin, A simple approach to integration of acceleration data for dynamic soil-structure interaction analysis. Soil Dyn. Earthq. Eng. **26**, 725–734 (2006)
13. A. Link, H.J. von Martens, Accelerometer identification using shock excitation. Measurement **35**, 191–199 (2004)
14. A.K. Chopra, *Dynamics of Structures: Theory and application to Earthquake engineering*, 2nd edn. (Prentice Hall, 2001)
15. Freescale Semiconductor, Accelerometer terminology guide. http://cache.freescale.com/files/sensors/doc/support_info/SENSORTERMSPG.pdf (2007)
16. W.H. Zhu, Velocity estimation by using position and acceleration sensors. IEEE Trans. Ind. Electron. **54**, 2706–2715 (2007)
17. D.M. Boore, Analog-to-digital conversion as a source of drifts in displacements derived from digital recordings of ground acceleration. Bull. Seismol. Soc. Am. **93**, 2017–2024 (2003)
18. T.S. Edwards, Effects of aliasing on numerical integration. Mech. Syst. Signal Process. **21**, 165–176 (2007)

Chapter 4
Stable PID Active Control of Building Structures

Abstract In this chapter, we analyze the stability of the active vibration control system for both linear and nonlinear structures. We give explicit sufficient conditions for choosing the PID gains. The theory conclusions are verified via numerical simulations and a two-story building prototype. These results give validation of our theory analysis.

Keywords Structural control · PID control · Stability

The building structures are vulnerable to natural and man-made hazards, which may result in financial, environmental, and human losses. It is essential to protect these structures, including the human occupants and nonstructural components from these threats. One approach to mitigate this undesirable behavior is to alter the dynamic characteristics of the building with respect to a given load, which can be achieved by adding control devices like dampers or actuators to the building [1].

Control device and controller design are the main focus of the traditional active vibration control systems [2, 3]. Since the force exerted by the earthquake and wind on the structures are very huge and uncertain, these large civil structures require a large amount of energy to control it. The structural control can be classified as passive control which does not require an external power source [4], and active control which uses sensors and active actuators to control the unwanted vibrations [5]. There are many active control devices designed for structural control applications [6]. The active mass damper (AMD) is the most popular actuator, which uses a mass without spring and dashpot [7]. In this chapter, we use AMD-type actuator for the active vibration control.

In order to achieve a good performance, it is essential to design an effective control strategy, which should be simple, robust, and fault tolerant. Many attempts have been made to introduce advanced controllers for the active vibration control of building structures. Instead of changing the structure stiffness, a pole-placement H_∞ control corresponding to a target damping ratio is proposed in [8]. In order to avoid the higher order problem in H_∞ control, the balanced truncation is applied in [9]. In [10], the genetic algorithm is used to determine the feedback control. There are

© The Author(s) 2016
W. Yu and S. Thenozhi, *Active Structural Control with Stable
Fuzzy PID Techniques*, SpringerBriefs in Applied Sciences and Technology,
DOI 10.1007/978-3-319-28025-7_4

several optimal control algorithms applied for the active vibration control of building structures, for example, filtered linear quadratic control (LQ) [11], linear quadratic regulator (LQR) [12], and linear quadratic Gaussian (LQG) [1]. All these controllers are model-based, which are complex and demands the exact model of the building structure. Some model-free controllers, such as sliding mode control (SMC) [13], neural network control [14], and fuzzy logic control [15] are still complex.

Proportional-integral-derivative (PID) control is widely used in industrial applications. Without model knowledge, PID control may be the best controller in real-time applications [16]. The great advantages of PID control over the others are that they are simple and have clear physical meanings. Although theory research in PID control algorithms is well established, it is still not well developed in structural vibration control. In [17], a simple proportional control is applied to reduce the building displacement due to wind excitation. In [18, 19], PD and PID controllers were used in the numerical simulations. In [20], a proportional-integral (PI) controller with an AMD is used to attenuate the structural motion due to earthquake. However, these control results are not satisfactory, because it is difficult to tune PID gains to guarantee good performances such as rise time, overshoot, settling time, and steady-state error [19]. Moreover, these works do not discuss the stability analysis of these active control systems.

While there is no doubt about the advances in the structural control field, there still exist some areas which need more exploration [21]. The active devices have the ability to add force onto the building structure. A poorly designed controller will lead to an undesirable control performance, which can even damage the building. So it is desired to study the stability of the closed-loop system. Only a few structural controllers such as H_∞ and SMC consider the stability in their design, whereas the other control strategies do not. However, these designs have concerned only the linear stiffness, since it represents a simple and efficient model at least for a small operational range. In practice, these building structures possess nonlinear behavior like the hysteresis phenomenon [22]. Also, there is a lack of experimental verification of these controllers. The practical implementation of a controller will be challenging if these issues were not addressed.

In this chapter, we use standard industrial PD and PID controllers for the active vibration control. The main contribution is that we give theory analysis of these PD/PID controllers. Both the linear and nonlinear cases for structural stiffness are considered in the analysis. Bouc–Wen model is used to model the nonlinear hysteresis phenomenon. The sufficient conditions for asymptotic stability are derived, which are simple and explicit. The controller gains can be decided directly from these conditions. Numerical simulations are given to compare with SMC. An active vibration control system for a six-story building structure equipped with an AMD is constructed for the experimental study. The experimental results via PD and PID controllers are discussed and the effectiveness of our theory results is demonstrated.

4.1 Stable PD Control

PD control may be the simplest controller for the structural vibration control system, which provides high robustness with respect to uncertainties. PD control has the following form:

$$\mathbf{u} = -K_p(\mathbf{x} - \mathbf{x}^d) - K_d(\dot{\mathbf{x}} - \dot{\mathbf{x}}^d) \tag{4.1}$$

where K_p and K_d are positive definite constant matrices, which correspond to the proportional and derivative gains, respectively, and $\mathbf{x}^d$ is the desired position. In active vibration control of building structures, the references are $\mathbf{x}^d = \dot{\mathbf{x}}^d = 0$, hence (4.1) becomes

$$\mathbf{u} = -K_p\mathbf{x} - K_d\dot{\mathbf{x}} \tag{4.2}$$

The aim of the controller design is to choose the suitable gains K_p and K_d in (4.2), such that the closed-loop system is stable. Without loss of generality, we use a six-story building structure as shown in Fig. 4.1.

When the structural parameters in (5.2) are completely known, i.e., there are no uncertainties and $\mathbf{f}_s$ is linear as in (2.8), the building structure is a linear determinant system. Many papers have used this model for the structure control design, such as PID control [18], H_2 control [9], and optimal control [12]. However, they did not discuss the stability problem.

Assuming $\mathbf{d} = \mathbf{0}$, the closed-loop system with the PD control in (4.2) is

$$M\ddot{\mathbf{x}} + C\dot{\mathbf{x}} + K\mathbf{x} + \mathbf{f}_e = \Gamma(-K_p\mathbf{x} - K_d\dot{\mathbf{x}}) \tag{4.3}$$

where $M = \begin{bmatrix} m_1 & 0 \\ 0 & m_2 \end{bmatrix} > 0$, $C = \begin{bmatrix} c_1 + c_2 & -c_2 \\ -c_2 & c_2 \end{bmatrix} > 0$, $K = \begin{bmatrix} k_1 + k_2 & -k_2 \\ -k_2 & k_2 \end{bmatrix} > 0$,

$\mathbf{x} = \begin{bmatrix} x_1 \\ x_2 \end{bmatrix}$, $\mathbf{f}_e = \begin{bmatrix} m_1\ddot{x}_g \\ m_2\ddot{x}_g \end{bmatrix}$, $K_p = \begin{bmatrix} k_{p1} & 0 \\ 0 & k_{p2} \end{bmatrix} > 0$, and $K_d = \begin{bmatrix} k_{d1} & 0 \\ 0 & k_{d2} \end{bmatrix} > 0$. The

damper is installed on the second floor, then $\Gamma = \begin{bmatrix} 0 & 0 \\ 0 & 1 \end{bmatrix}$. Now we are in a position to study the system represented in (4.3) using linear techniques. Equation (4.3) can be written in the state-space form

$$\dot{\mathbf{z}} = A_{cl}\mathbf{z} + \mathbf{f}_{cl} \tag{4.4}$$

Fig. 4.1 PD/PID control for a two-story building

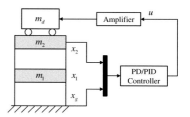

where $\mathbf{z} = \begin{bmatrix} \mathbf{x} \\ \dot{\mathbf{x}} \end{bmatrix} \in \Re^4$, $A_{cl} = \begin{bmatrix} 0_{2\times2} & I_{2\times2} \\ -M^{-1}\left(K + \Gamma K_p\right) & -M^{-1}\left(C + \Gamma K_d\right) \end{bmatrix} \in \Re^{4\times4}$,

and $\mathbf{f}_{cl} = \begin{bmatrix} 0_{1\times2} & -\mathbf{f}_e^T \end{bmatrix}^T \in \Re^4$.

The stability of the closed-loop system in (4.4) depends on the system matrix A_{cl}. Its characteristic polynomial is

$$\det(sI - A_{cl}) = s^4 + a_1 s^3 + a_2 s^2 + a_3 s + a_4 \tag{4.5}$$

where

$$a_1 = \frac{1}{m_1}(c_1 + c_2) + \frac{1}{m_2}(c_2 + k_{d2})$$

$$a_2 = \frac{1}{m_1 m_2}\left(c_1 k_{d2} + c_2 k_{d2} + m_1 k_{p2} + c_1 c_2 + k_1 m_2 + k_2 m_1 + k_2 m_2\right)$$

$$a_3 = \frac{1}{m_1 m_2}\left(k_1 k_{d2} + k_2 k_{d2} + c_1 k_{p2} + c_2 k_{p2} + c_1 k_2 + c_2 k_1\right)$$

$$a_4 = \frac{1}{m_1 m_2}\left(k_1 k_{p2} + k_2 k_{p2} + k_1 k_2\right) \tag{4.6}$$

Using Lienard–Chipart criterion, the closed-loop system A_{cl} is stable if and only if

$$a_i > 0, i = 1, 2, 3, 4 \text{ and } a_1 a_2 a_3 - a_1^2 a_4 - a_3^2 > 0 \tag{4.7}$$

Now the designer can directly choose the controller gains, which can satisfy the five inequalities given by (4.7).

In practice, the parameters of the building structure are partly known and the structure model might have nonlinearities such as the hysteresis phenomenon. It is convenient to express (4.3) as

$$M\ddot{\mathbf{x}} + C\dot{\mathbf{x}} + \mathbf{f} = \Gamma\mathbf{u} \tag{4.8}$$

where

$$\mathbf{f} = \mathbf{f}_s + \mathbf{f}_e + \mathbf{d} \tag{4.9}$$

The building structure with the PD control in (4.2) can be now written as

$$M\ddot{\mathbf{x}} + C\dot{\mathbf{x}} + \mathbf{f} = -\Gamma\left(K_p\mathbf{x} + K_d\dot{\mathbf{x}}\right) \tag{4.10}$$

Since (4.10) is a nonlinear system and M, C, and $\mathbf{f}$ are unknown, Routh–Hurwitz stability criterion in (4.5) cannot be applied here. The following theorem gives the stability analysis of the PD control in (4.2). In order to simplify the proof, we first consider $\Gamma_{n\times n} = I_{n\times n}$, i.e., each floor has an actuator installed on it.

Theorem 4.1.1 *Consider the structural system as (4.8) controlled by the PD con-troller as (4.2), the closed-loop system as (4.10) is stable, provided that the control gains satisfy*

$$K_p > 0, \quad K_d > 0 \tag{4.11}$$

Here $K > 0$ means K is a positive definite matrix, i.e., with any vector $\mathbf{x}$, $\mathbf{x}^T K \mathbf{x} > 0$, all of its eigenvalues are positive. The derivative of the regulation error $\mathbf{x}$ converges to the residual set

$$D_{\dot{\mathbf{x}}} = \left\{ \dot{\mathbf{x}} \mid \|\dot{\mathbf{x}}\|_Q^2 \le \bar{\mu}_{\mathbf{f}} \right\} \tag{4.12}$$

where $\bar{\mu}_{\mathbf{f}} \ge \mathbf{f}^T \Lambda_{\mathbf{f}}^{-1} \mathbf{f}$ and $C > \Lambda_{\mathbf{f}} > 0$.

Proof We select the system's energy as the Lyapunov candidate V:

$$V = \frac{1}{2} \dot{\mathbf{x}}^T M \dot{\mathbf{x}} + \frac{1}{2} \mathbf{x}^T K_p \mathbf{x} \tag{4.13}$$

The first term of (4.13) represents the kinetic energy and the second term is the virtual elastic potential energy. Since M and K_p are positive definite matrices, $V \ge 0$. The derivative of (4.13) is

$$
\begin{aligned}
\dot{V} &= \dot{\mathbf{x}}^T M \ddot{\mathbf{x}} + \dot{\mathbf{x}}^T K_p \mathbf{x} \\
&= \dot{\mathbf{x}}^T \left(-C\dot{\mathbf{x}} - \mathbf{f} - K_p \mathbf{x} - K_d \dot{\mathbf{x}} \right) + \dot{\mathbf{x}}^T K_p \mathbf{x} \\
&= -\dot{\mathbf{x}}^T (C + K_d) \dot{\mathbf{x}} - \dot{\mathbf{x}}^T \mathbf{f}
\end{aligned}
\tag{4.14}
$$

Using the matrix inequality

$$X^T Y + Y^T X \le X^T \Lambda X + Y^T \Lambda^{-1} Y \tag{4.15}$$

which is valid for any $X, Y \in \Re^{n \times m}$ and any $0 < \Lambda = \Lambda^T \in \Re^{n \times n}$, we can write the scalar variable $\dot{\mathbf{x}}^T \mathbf{f}$ as

$$\dot{\mathbf{x}}^T \mathbf{f} = \frac{1}{2} \dot{\mathbf{x}}^T \mathbf{f} + \frac{1}{2} \mathbf{f}^T \dot{\mathbf{x}} \le \dot{\mathbf{x}}^T \Lambda_{\mathbf{f}} \dot{\mathbf{x}} + \mathbf{f}^T \Lambda_{\mathbf{f}}^{-1} \mathbf{f} \tag{4.16}$$

where $\Lambda_{\mathbf{f}}$ is any positive definite matrix. In this theorem, we select $\Lambda_{\mathbf{f}}$ as

$$C > \Lambda_{\mathbf{f}} > 0 \tag{4.17}$$

So

$$\dot{V} \le -\dot{\mathbf{x}}^T (C + K_d - \Lambda_{\mathbf{f}}) \dot{\mathbf{x}} + \mathbf{f}^T \Lambda_{\mathbf{f}}^{-1} \mathbf{f} \tag{4.18}$$

If we choose $K_d > 0$, then

$$\dot{V} \le -\dot{\mathbf{x}}^T Q \dot{\mathbf{x}} + \bar{\mu}_{\mathbf{f}} \le -\lambda_m (Q) \|\dot{\mathbf{x}}\|^2 + \mathbf{f}^T \Lambda_{\mathbf{f}}^{-1} \mathbf{f} \tag{4.19}$$

where $Q = K_d + C - \Lambda_f > 0$. V is therefore an ISS-Lyapunov function. Using Theorem 1 from [23], the boundedness of $\mathbf{f}^T \Lambda_f^{-1} \mathbf{f} \leq \bar{\mu}_f$ implies that the regulation error $\|\dot{\mathbf{x}}\|$ is bounded. It is noted that when

$$\|\dot{\mathbf{x}}\|_Q^2 > \bar{\mu}_f, \quad \forall t \in [0, T] \tag{4.20}$$

$\dot{V} < 0$. Now we prove that the total time during which $\|\dot{\mathbf{x}}\|_Q^2 > \bar{\mu}_f$ is finite. Let T_k denotes the time interval during which $\|\dot{\mathbf{x}}\|_Q^2 > \bar{\mu}_f$.

(1) If only finite times that $\|\dot{\mathbf{x}}\|_Q^2 > \bar{\mu}_f$ stay outside the circle of radius $\bar{\mu}_f$ (and then reenter), $\|\dot{\mathbf{x}}\|_Q^2 > \bar{\mu}_f$ will eventually stay inside of this circle.
(2) If $\|\dot{\mathbf{x}}\|_Q^2 > \bar{\mu}_f$ leave the circle infinite times, since the total time $\|\dot{\mathbf{x}}\|_Q^2 > \bar{\mu}_f$ leave the circle is finite, then

$$\sum_{k=1}^{\infty} T_k < \infty, \quad \lim_{k \to \infty} T_k = 0 \tag{4.21}$$

So $\|\dot{\mathbf{x}}\|_Q^2$ is bounded via an invariant set argument. From (4.19) $\|\dot{\mathbf{x}}\|$ is also bounded. Let $\|\dot{\mathbf{x}}\|_Q^2$ denotes the largest tracking error during the T_k interval. Then (4.21) and bounded $\|\dot{\mathbf{x}}\|_Q^2$ imply that

$$\lim_{k \to \infty} \left[\|\dot{\mathbf{x}}\|_Q^2 - \bar{\mu}_f \right] = 0 \tag{4.22}$$

So $\|\dot{\mathbf{x}}\|_Q^2$ will converge to $\bar{\mu}_f$, and hence (4.12) is achieved.

Since $V \geq 0$, V decreases until $\|\dot{\mathbf{x}}\|_Q^2 \leq \bar{\mu}_f$. Total time of $\|\dot{\mathbf{x}}\|_Q^2 > \bar{\mu}_f$ being finite means that $V = \frac{1}{2}\dot{\mathbf{x}}^T M \dot{\mathbf{x}} + \frac{1}{2}\mathbf{x}^T K_p \mathbf{x}$ is bounded; hence, the regulation error $\dot{\mathbf{x}}$ is bounded.

It is well known that the regulation error becomes smaller while increasing the gain K_d. The cost of large K_d is that the transient performance becomes slow. Only when $K_d \to \infty$, the regulation error converges to zero [24]. However, it would seem better to use a smaller K_d if the system contains high-frequency noise signals.

4.2   Stable PID Control

From the above section it is clear that any increase in the derivative gain K_d can decrease the regulation error, but causes a slow response. In the control viewpoint, the regulation error is removed by introducing an integral component to the PD control, i.e., modify the PD control into PID control. The PID control law can be expressed as

$$\mathbf{u} = -K_p(\mathbf{x} - \mathbf{x}^d) - K_i \int_0^t (\mathbf{x} - \mathbf{x}^d)d\tau - K_d(\dot{\mathbf{x}} - \dot{\mathbf{x}}^d) \tag{4.23}$$

where $K_i > 0$ correspond to the integration gain. For the regulation case $x^d = \dot{x}^d = 0$, (4.23) becomes

$$\mathbf{u} = -K_p\mathbf{x} - K_i \int_0^t \mathbf{x} d\tau - K_d\dot{\mathbf{x}} \tag{4.24}$$

In order to analyze the stability of PID controller, (4.24) is expressed by

$$\mathbf{u} = -K_p\mathbf{x} - K_d\dot{\mathbf{x}} - \xi$$
$$\xi = K_i\mathbf{x}, \quad \xi(0) = \mathbf{0} \tag{4.25}$$

Now substituting (4.25) into (5.7), the closed-loop system can be written as

$$M\ddot{\mathbf{x}} + C\dot{\mathbf{x}} + \mathbf{f} = -K_p\mathbf{x} - K_d\dot{\mathbf{x}} - \xi \tag{4.26}$$

In matrix form, the closed-loop system is

$$\frac{d}{dt}\begin{bmatrix} \xi \\ \mathbf{x} \\ \dot{\mathbf{x}} \end{bmatrix} = \begin{bmatrix} K_i\mathbf{x} \\ \dot{\mathbf{x}} \\ -M^{-1}\left(C\dot{\mathbf{x}} + \mathbf{f} + K_p\mathbf{x} + K_d\dot{\mathbf{x}} + \xi\right) \end{bmatrix} \tag{4.27}$$

The equilibrium of (4.27) is $[\xi, \mathbf{x}, \dot{\mathbf{x}}] = [\xi^*, \mathbf{0}, \mathbf{0}]$. Since at equilibrium point $\mathbf{x} = \mathbf{0}$ and $\dot{\mathbf{x}} = \mathbf{0}$, the equilibrium is $[\mathbf{f}(0), \mathbf{0}, \mathbf{0}]$. In order to move the equilibrium to origin, we define

$$\xi = \xi - \mathbf{f}(0) \tag{4.28}$$

The final closed-loop equation becomes

$$M\ddot{\mathbf{x}} + C\dot{\mathbf{x}} + \mathbf{f} = -K_p\mathbf{x} - K_d\dot{\mathbf{x}} - \xi + \mathbf{f}(0)$$
$$\xi = K_i\mathbf{x} \tag{4.29}$$

In order to analyze the stability of (4.29), we first give the following properties.

P1. The positive definite matrix M satisfies the following condition:

$$0 < \lambda_m(M) \le \|M\| \le \lambda_M(M) \le \bar{m} \tag{4.30}$$

where $\lambda_m(M)$ and $\lambda_M(M)$ are the minimum and maximum eigenvalues of the matrix M, respectively, and $\bar{m} > 0$ is the upper bound.

P2. The term $\mathbf{f}$ is Lipschitz over $\tilde{x}$ and $\tilde{y}$

$$\|\mathbf{f}(\tilde{x}) - \mathbf{f}(\tilde{y})\| \le k_{\mathbf{f}} \|\tilde{x} - \tilde{y}\| \tag{4.31}$$

Most of the uncertainties are first-order continuous functions. Since $\mathbf{f}_s$, $\mathbf{f}_e$, and $\mathbf{d}$ are first-order continuous functions and satisfy Lipschitz condition, **P2** can be established using (5.8). Now we calculate the lower bound of $\int \mathbf{f} \, dx$.

$$\int_0^t \mathbf{f} dx = \int_0^t \mathbf{f}_s dx + \int_0^t \mathbf{f}_e dx + \int_0^t \mathbf{d} dx \qquad (4.32)$$

We define the lower bound of $\int_0^t \mathbf{f}_s dx$ is $-\bar{f}_s$ and for $\int_0^t \mathbf{d}\, dx$ is $-\bar{d}$. Compared with $\mathbf{f}_s$ and $\mathbf{d}$, $\mathbf{f}_e$ is much bigger in the case of earthquake. We define the lower bound of $\int_0^t \mathbf{f}_e dx$ is $-\bar{f}_e$. Finally, the lower bound $k_\mathbf{f}$ is

$$k_\mathbf{f} = -\bar{f}_s - \bar{f}_e - \bar{d} \qquad (4.33)$$

The following theorem gives the stability analysis of PID controller (4.25).

Theorem 4.2.1 *Consider the structural system as (5.7) controlled by the PID controller as (4.25), the closed-loop system as (4.29) is asymptotically stable at the equilibrium $[\xi - \mathbf{f}(0), \mathbf{x}, \dot{\mathbf{x}}]^T = 0$, provided that the control gains satisfy*

$$\begin{aligned}
\lambda_m(K_p) &\geq \tfrac{3}{2}[k_\mathbf{f} + k_c] \\
\lambda_M(K_i) &\leq \phi \frac{\lambda_m(K_p)}{\lambda_M(M)} \\
\lambda_m(K_d) &\geq \phi\left[1 + \frac{k_c}{\lambda_M(M)}\right] - \lambda_m(C)
\end{aligned} \qquad (4.34)$$

where $\phi = \sqrt{\tfrac{1}{3}\lambda_m(M)\lambda_m(K_p)}$.

Proof Here, the Lyapunov function is defined as

$$V = \frac{1}{2}\dot{\mathbf{x}}^T M\dot{\mathbf{x}} + \frac{1}{2}\mathbf{x}^T K_p\mathbf{x} + \frac{\alpha}{2}\xi^T K_i^{-1}\xi + \mathbf{x}^T \xi + \alpha\mathbf{x}^T M\dot{\mathbf{x}} + \frac{\alpha}{2}\mathbf{x}^T K_d\mathbf{x} + \int_0^t \mathbf{f} dx - k_\mathbf{f} \qquad (4.35)$$

where $k_\mathbf{f}$ is defined in (4.35) such that $V(0) = 0$. In order to show that $V \geq 0$, it is separated into three parts, such that $V = \sum_{i=1}^3 V_i$

$$V_1 = \frac{1}{6}\mathbf{x}^T K_p\mathbf{x} + \frac{\alpha}{2}\mathbf{x}^T K_d\mathbf{x} + \int_0^t \mathbf{f} dx - k_\mathbf{f} \geq 0 \qquad (4.36)$$

$$\begin{aligned}
V_2 &= \tfrac{1}{6}\mathbf{x}^T K_p\mathbf{x} + \tfrac{\alpha}{2}\xi^T K_i^{-1}\xi + \mathbf{x}^T\xi \\
&\geq \tfrac{1}{2}\tfrac{1}{6}\lambda_m(K_p)\|\mathbf{x}\|^2 + \tfrac{\alpha\lambda_m(K_i^{-1})}{2}\|\xi\|^2 - \|\mathbf{x}\|\,\|\xi\|
\end{aligned} \qquad (4.37)$$

When $\alpha \geq \frac{3}{\lambda_m(K_i^{-1})\lambda_m(K_p)}$,

$$V_2 \geq \frac{1}{2}\left(\sqrt{\frac{\lambda_m(K_p)}{3}}\|\mathbf{x}\| - \sqrt{\frac{3}{\lambda_m(K_p)}}\|\xi\|\right)^2 \geq 0 \qquad (4.38)$$

and

$$V_3 = \frac{1}{6}\mathbf{x}^T K_p\mathbf{x} + \frac{1}{2}\dot{\mathbf{x}}^T M\dot{\mathbf{x}} + \alpha\mathbf{x}^T M\dot{\mathbf{x}} \qquad (4.39)$$

$$\tilde{y}^T A \tilde{x} \geq \|\tilde{y}\| \, \|A\tilde{x}\| \geq \|\tilde{y}\| \, \|A\| \, \|\tilde{x}\| \geq |\lambda_M(A)| \, \|\tilde{y}\| \, \|\tilde{x}\| \tag{4.40}$$

when $\alpha \leq \dfrac{\sqrt{\frac{1}{3}\lambda_m(M)\lambda_m(K_p)}}{\lambda_M(M)}$

$$V_3 \geq \frac{1}{2}\left(\frac{1}{3}\lambda_m(K_p)\|\mathbf{x}\|^2 + \lambda_m(M)\|\dot{\mathbf{x}}\|^2 + 2\alpha\lambda_M(M)\|\mathbf{x}\|\,\|\dot{\mathbf{x}}\|\right)$$
$$= \frac{1}{2}\left(\sqrt{\frac{\lambda_m(K_p)}{3}}\|\mathbf{x}\| + \sqrt{\lambda_m(M)}\|\dot{\mathbf{x}}\|\right)^2 \geq 0 \tag{4.41}$$

If

$$\sqrt{\frac{1}{3}\lambda_m(K_i^{-1})\lambda_m^{\frac{3}{2}}(K_p)\lambda_m^{\frac{1}{2}}(M)} \geq \lambda_M(M) \tag{4.42}$$

there exists

$$\frac{\sqrt{\frac{1}{3}\lambda_m(M)\lambda_m(K_p)}}{\lambda_M(M)} \geq \alpha \geq \frac{3}{\lambda_m(K_i^{-1})\lambda_m(K_p)} \tag{4.43}$$

The derivative of (4.35) is

$$\begin{aligned}\dot{V} &= \dot{\mathbf{x}}^T M \ddot{\mathbf{x}} + \dot{\mathbf{x}}^T K_p \mathbf{x} + \alpha \xi^T K_i^{-1} \xi + \dot{\mathbf{x}}^T \xi + \mathbf{x}^T \xi + \alpha \dot{\mathbf{x}}^T M \dot{\mathbf{x}} \\ &\quad + \alpha \mathbf{x}^T M \ddot{\mathbf{x}} + \alpha \dot{\mathbf{x}}^T K_d \mathbf{x} + \dot{\mathbf{x}}^T \mathbf{f} \\ &= \dot{\mathbf{x}}^T \left[-C\dot{\mathbf{x}} - \mathbf{f} - K_p \mathbf{x} - K_d \dot{\mathbf{x}} - \xi + \mathbf{f}(0)\right] + \dot{\mathbf{x}}^T K_p \mathbf{x} \\ &\quad + \alpha \xi^T K_i^{-1} \xi + \dot{\mathbf{x}}^T, + \mathbf{x}^T \xi \\ &\quad + \alpha \dot{\mathbf{x}}^T M \dot{\mathbf{x}} + \alpha \mathbf{x}^T \left[-C\dot{\mathbf{x}} - \mathbf{f} - K_p \mathbf{x} - K_d \dot{\mathbf{x}} - \xi + \mathbf{f}(0)\right] + \alpha \mathbf{x}^T K_d \dot{\mathbf{x}} + \dot{\mathbf{x}}^T \mathbf{f}\end{aligned} \tag{4.44}$$

From (4.31)

$$\alpha \mathbf{x}^T [\mathbf{f}(0) - \mathbf{f}] \leq \alpha k_{\mathbf{f}} \|\mathbf{x}\|^2 \tag{4.45}$$

Using (4.15) we can write

$$-\alpha \mathbf{x}^T C \dot{\mathbf{x}} \leq \alpha k_c \left(\mathbf{x}^T \mathbf{x} + \dot{\mathbf{x}}^T \dot{\mathbf{x}}\right) \tag{4.46}$$

where $\|C\| \leq k_c$.

Since $\xi = K_i \mathbf{x}$, $\xi^T K_i^{-1} \xi$ becomes $\alpha \mathbf{x}^T \xi$ and $\mathbf{x}^T \xi$ becomes $\mathbf{x}^T K_i \mathbf{x}$, then

$$\dot{V} = -\dot{\mathbf{x}}^T [C + K_d - \alpha M - \alpha k_c]\dot{\mathbf{x}} - \mathbf{x}^T \left[\alpha K_p - K_i - \alpha k_{\mathbf{f}} - \alpha k_c\right]\mathbf{x} \tag{4.47}$$

Using (5.40), (4.47) becomes

$$\begin{aligned}\dot{V} &\leq -\dot{\mathbf{x}}^T [\lambda_m(C) + \lambda_m(K_d) - \alpha\lambda_M(M) - \alpha k_c]\dot{\mathbf{x}} \\ &\quad - \mathbf{x}^T \left[\alpha\lambda_m(K_p) - \lambda_M(K_i) - \alpha k_{\mathbf{f}} - \alpha k_c\right]\mathbf{x}\end{aligned} \tag{4.48}$$

If $\lambda_m(C) + \lambda_m(K_d) \geq \alpha [\lambda_M(M) + k_c]$ and $\lambda_m(K_p) \geq \frac{1}{\alpha}\lambda_M(K_i) + k_f + k_c$, then $\dot{V} \leq 0$, $\|\mathbf{x}\|$ decreases. From (5.52) and $\lambda_m(K_i^{-1}) = \frac{1}{\lambda_M(K_i)}$, if

$$\lambda_m(K_d) \geq \sqrt{\frac{1}{3}\lambda_m(M)\lambda_m(K_p)}\left[1 + \frac{k_c}{\lambda_M(M)}\right] - \lambda_m(C)$$
$$\lambda_m(K_p) \geq \frac{3}{2}[k_f + k_c] \tag{4.49}$$

then (4.34) is established.

Finally, we prove the asymptotic stability of the closed-loop system as (4.29). There exists a ball Σ of radius $\rho > 0$ centered at the origin of the state-space on which $\dot{V} \leq 0$. The origin of the closed-loop equation as (4.29) is a stable equilibrium. Since the closed-loop equation is autonomous, we use La Salle's theorem. Define Ω as

$$\Omega = \left\{\bar{\mathbf{z}}(t) = \left[\mathbf{x}^T, \dot{\mathbf{x}}^T, \xi^T\right]^T \in \mathfrak{R}^{3n} : \dot{V} = 0\right\}$$
$$= \left\{\xi \in \mathfrak{R}^n, \mathbf{x} = \mathbf{0} \in \mathfrak{R}^n, \dot{\mathbf{x}} = 0 \in \mathfrak{R}^n\right\} \tag{4.50}$$

From (4.44), $\dot{V} = 0$ if and only if $\mathbf{x} = \dot{\mathbf{x}} = \mathbf{0}$. For a solution $\bar{\mathbf{z}}(t)$ to belong to Ω for all $t \geq 0$, it is necessary and sufficient that $\mathbf{x} = \dot{\mathbf{x}} = \mathbf{0}$ for all $t \geq 0$. Therefore, it must also hold that $\ddot{\mathbf{x}} = \mathbf{0}$ for all $t \geq 0$. We conclude that from the closed-loop system as (4.29), if $\bar{\mathbf{z}}(t) \in \Omega$ for all $t \geq 0$, then $\mathbf{f}(\mathbf{x}) = \mathbf{f}(0) = \xi + \mathbf{f}(0)$ and $\dot{\xi} = \mathbf{0}$. It implies that $\xi = \mathbf{0}$ for all $t \geq 0$. So $\bar{\mathbf{z}}(t) = \mathbf{0}$ is the only initial condition in Ω for which $\bar{\mathbf{z}}(t) \in \Omega$ for all $t \geq 0$. We conclude from the above discussions that the origin of the closed-loop system as (4.29) is asymptotically stable. It establishes the stability of the proposed controller, in the sense that the domain of attraction can be enlarged with a suitable choice of the gains. Namely, increasing K_p the basin of attraction will grow.

Remark 4.2.1 Since the stiffness of the building structure has hysteresis property, the hysteresis output depends on both the instantaneous and the history of the deformation. This deformation before applying the force (loading) and after removing the force (unloading) is not the same, i.e., the equilibrium position before the earthquake and after the vibration dies out is not the same. After the earthquake, the stable point is moved. This corresponds to the term $\mathbf{f}(0)$. So we cannot conclude that the closed-loop system is globally stable.

It is well known that, in the absence of the uncertainties and external force, $\mathbf{f} = \mathbf{0}$, the PD control as (5.6) with any positive gains can drive the closed-loop system asymptotically stable. The main objective of the integral action can be regarded to cancel $\mathbf{f}$. In order to decrease integral gain, an estimated $\mathbf{f}$ is applied to the PID control as (5.36). The PID control with an approximate force compensation $\hat{\mathbf{f}}$ is

$$\mathbf{u} = -K_p\mathbf{x} - K_d\dot{\mathbf{x}} - \xi + \hat{\mathbf{f}}, \quad \xi = K_i\mathbf{x} \tag{4.51}$$

The above theorem is also applicable for the PID controllers with an approximate $\mathbf{f}$ compensation as in (4.51). The condition for PID gains in (4.34) becomes $\lambda_m \left(K_p \right) \geq \frac{3}{2} \left[\tilde{k}_f + k_c \right]$ and $\lambda_M \left(K_i \right) \leq \frac{3\phi}{2} \frac{\tilde{k}_f + k_c}{\lambda_M (M)}$, $\tilde{k}_f \ll k_f$.

If the number of dampers installed on the buildings is less than the number of the building floors (n), then the resulting system is termed as under-actuated system. In this case, the location matrix Γ should be included along with the gain matrices. In our experiment, there is only one damper installed (second floor) on the structure. The PID controller becomes

$$\Gamma \mathbf{u} = \begin{bmatrix} 0 & 0 \\ 0 & 1 \end{bmatrix} \left\{ - \begin{bmatrix} k_{p1} & 0 \\ 0 & k_{p2} \end{bmatrix} \begin{bmatrix} x_1 \\ x_2 \end{bmatrix} - \begin{bmatrix} k_{i1} & 0 \\ 0 & k_{i2} \end{bmatrix} \begin{bmatrix} \int_0^t x_1 d\tau \\ \int_0^t x_2 d\tau \end{bmatrix} - \begin{bmatrix} k_{d1} & 0 \\ 0 & k_{d2} \end{bmatrix} \begin{bmatrix} \dot{x}_1 \\ \dot{x}_2 \end{bmatrix} \right\}$$
(4.52)

$$\Gamma \mathbf{u} = \begin{bmatrix} 0 \\ -k_{p2}x_2 - k_{i2} \int_0^t x_2 d\tau - k_{d2}\dot{x}_2 \end{bmatrix}$$
(4.53)

where the scalars k_{p2}, k_{i2}, and k_{d2} are the proportional, integral, and derivative gains, respectively. In this case (5.44) becomes

$$\begin{aligned} k_{p2} &\geq \frac{3}{2} [k_f + k_c] \\ k_{i2} &\leq \tilde{\phi} \frac{\min\{k_{p2}\}}{\lambda_M (M)} \\ k_{d2} &\geq \tilde{\phi} \left[1 + \frac{k_c}{\lambda_M (M)} \right] - \lambda_m (C) \end{aligned}$$
(4.54)

where $\tilde{\phi} = \sqrt{\frac{1}{3}\lambda_m (M) \min\{k_{p2}\}}$.

Remark 4.2.2 The PID tuning methods are different for the system with and without prior knowledge. If the system parameters are unknown, then auto-tuning techniques are employed to choose the gains either online or offline. These techniques are broadly classified into direct and indirect methods [25]. In direct method, the closed-loop response of the system is observed and the controller gains are tuned directly based on the past experience and heuristic rules. In the case of indirect method, the structure parameters are identified first from the measured output and based on these identified parameters the controller is then tuned to achieve a desired system dynamics. This chapter provides a tuning method that ensures a stable closed-loop performance. For that purpose, the structural parameters $\lambda_M (M)$, $\lambda_m(C)$, k_f, and k_c are determined from the identified parameters.

Remark 4.2.3 The PID control as (4.23) does not need exact information about the building structure as (4.3). It uses only the displacements of the building and upper bound estimation of the building parameters. If the actual control force to the building structure satisfies (5.44), the closed-loop system is stable, and this condition is easy to be satisfied from the above remark. So we does not require the theory force (4.23) to match actual control force for PID structure control. However, in many cases the actual control force cannot reach the theory force as (4.23) due to the actuator limitations, which causes saturation:

$$u_{\text{real}} = sat\left[u_{theory}\right] = \begin{cases} u_{theory} & \text{if } \left\|u_{theory}\right\| < \nu_{\max} \\ \nu_{\max} & \text{if } \left\|u_{theory}\right\| \geq \nu_{\max} \end{cases} \qquad (4.55)$$

where u_{theory} is the theory force, u_{real} is the actual control force, and $\nu_{\max}$ is the maximum torque of the AMD actuator. Now the linear PID controller becomes nonlinear PID. The asymptotic stability of Theorem 2 becomes stable as Theorem 1, see [26].

4.3 Simulations and Experimental Results

4.3.1 Numerical Simulations

Consider the system described by (2.6) with linear stiffness has the following set of parameters: the matrix M is $m_1 = 3.3$ and $m_2 = 6.1$ kg, C is given by $c_1 = 2.5$ and $c_2 = 1.4$ N s / m, and K is given by $k_1 = 4080$ and $k_2 = 4260$ N / m. These parameters are obtained by identifying the six-story lab prototype.

We compare the performances of PD, PID, and SMC. Like PID controller, SMC is a popular robust controller which is often seen in the structural vibration control applications [21]. A switching control law is used to drive the state trajectory onto a prespecified surface. In the case of structural vibration control, this surface corresponds to a desired system dynamics.

A general class of discontinuous structural control is defined by the following relationships [27]:

$$\mathbf{u} = \mathbf{u}_{eq} - \eta \text{sign}(\boldsymbol{\sigma}) = \begin{cases} -\eta & \text{if } \sigma > 0 \\ 0 & \text{if } \sigma = 0 , \\ \eta & \text{if } \sigma < 0 \end{cases} \quad \eta > 0 \qquad (4.56)$$

where the linear term $\mathbf{u}_{eq}$ is the equivalent control force, $\boldsymbol{\sigma}$ is the sliding surface, and $\text{sign}(\boldsymbol{\sigma}) = \left[\text{sign}(\sigma_1), \dots, \text{sign}(\sigma_{2n})\right]^T$. The sliding surface can be a function of the the regulation error, and then $\sigma = \left[\mathbf{x}^T, \dot{\mathbf{x}}^T\right]^T = \mathbf{z} \in \Re^{2n}$. The equivalent control can be estimated using a low-pass filter or neglected, if the system parameters are unknown.

If the PD control as (4.2) has $k_{p2} = 350$ and $k_{d2} = 45$, they satisfy the condition as (4.7); hence, the closed loop is stable. It is easy to verify that the closed-loop system with PID control ($A_{cl} \in \Re^{5 \times 5}$, $k_{p2} = 350$, $k_{i2} = 2200$, and $k_{d2} = 45$) is stable.

If the SMC switching gain η is greater than the system uncertainty bound, then the $\sigma = \mathbf{z}$ converges to zero. We consider the switching gain of SMC $\eta = 1.3$.

Here the structure is excited by a step input and the corresponding vibration response is reduced by applying the above controllers. The control objective is to bring the structural vibration as close to zero as possible. The control signal is directly applied as the force without applying any constraint by neglecting the damper dynamics.

Figure 4.2 shows the time response of the second floor displacement for both controlled and uncontrolled cases; the unite is centimeter. It shows that all the three controllers reduce the structure motion. The PD controller reduces the structure oscillations but has a big steady-state error. This error can be reduced by introducing an integral term, hence a PID controller, which can achieve a zero steady-state error. The control signals are shown in Figs. 4.3, 4.4 and 4.5; the unites are volt.

The performance of the SMC lies between the PD and PID controllers but its control signal has many high-frequency switching, which may not be acceptable for some mechanical dampers. However, in practice, the active system cannot achieve this much attenuation due to the actuator limitations.

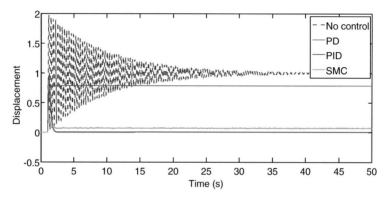

Fig. 4.2 The displacements of the second floor using PD, PID, and SMC controls

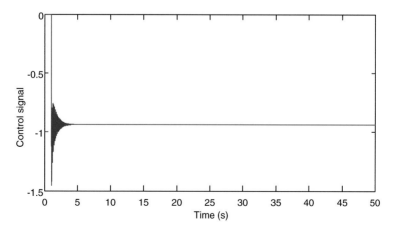

Fig. 4.3 Control signal of PD control

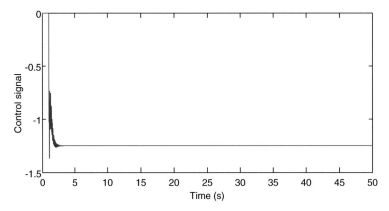

Fig. 4.4 Control signal of PID control

Fig. 4.5 Control signal of SMC control

4.3.2   Experimental Results

To illustrate the theory analysis results, a six-story building prototype is constructed which is mounted on a shaking table, see Fig. 4.6. The building structure is constructed of aluminum. The shaking table is actuated using a hydraulic control system (FEEDBACK EHS 160), which is used to generate earthquake signals. The AMD is a linear servo actuator (STB1108, Copley Controls Corp.), which is mounted on the second floor. The moving mass of the damper weighs 5 % (0.45 % kg) of the total building mass. The linear servo mechanism is driven by a digital servo drive (Accelnet Micro Panel, Copley Controls Corp). ServoToGo II I/O board is used for the data acquisition purpose.

The PD/PID control needs the structure position and velocity data. During the seismic excitation, the reference where the displacement and velocity sensors are attached will also move, and as a result the absolute value of the above parameters cannot be sensed. Alternatively, accelerometers can provide inexpensive and reliable

Fig. 4.6 Seven-story
building prototype with the
shaking table

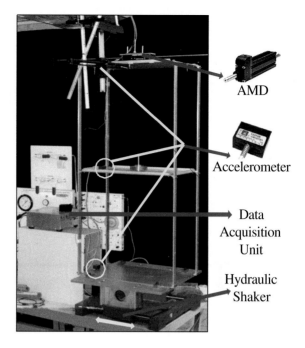

AMD

Accelerometer

Data
Acquisition
Unit

Hydraulic
Shaker

measurement of the acceleration at strategic points on the structure. Three accelerometers (Summit Instruments 13203B) were used to measure the absolute accelerations
on the ground and each floor. The ground acceleration is then subtracted from the
each floor acceleration to get the relative floor movement. The relative velocity and
position data are then estimated using the numerical integrator proposed in [21].

The position estimation with respect to the Loma Prieta East-West earthquake signal is shown in Fig. 4.7. The effect of the proposed numerical integrator on frequency
characteristics is studied by plotting Fourier spectra. A sinusoidal signal composed
with 6, 7, and 8 Hz is used here to excite the linear actuator. The linear actuator
has a position and acceleration sensor. The acceleration of the actuator is measured,
which is then integrated twice to obtain the position estimation. The FFT diagram of
the measured and estimated position is generated, see Fig. 4.8. As can be seen from
the figure, the frequency information is not affected except in the low-frequency
range. This low-frequency error is caused due to the presence of bias and noise in
the accelerometer output.

The control programs were operated in Windows XP with Matlab 6.5/Simulink.
All the control actions were employed at a sampling frequency of 1.0 kHz. The control signal generated by the control algorithm is fed as voltage input to the amplifier.
The current control loop is used to control the AMD operation. The amplifier con-
verts its voltage input to a respective current output with a gain of 0.5. The AMD
has a force constant of 6.26 N/A or 3.13 N/V.

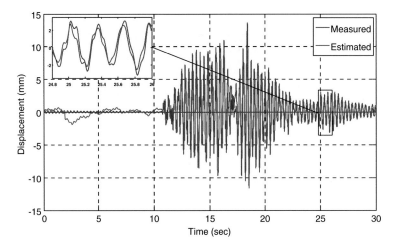

Fig. 4.7 Comparison of the measured and estimated position data

Fig. 4.8 Comparison of the measured and estimated position data using Fourier spectra

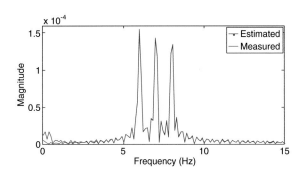

Now, we show the procedure for selecting the gains for a stable operation. The theorems in this chapter give sufficient conditions for the minimal values of the proportional and derivative gains and maximal values of the integral gains. In order to do a fair comparison, both the PD and PID controllers use the same proportional and derivative gains. We first design the PID controller based on the identified parameters of the six-story lab prototype. The following set of parameters were used for the control design: $\lambda_M(M) = 6.1$, $\lambda_m(C) = 0.6$, $k_f = 365$, and $k_c = 5.8$. Applying these values in Theorem 2, we get

$$\lambda_m(K_p) \geq 556, \quad \lambda_M(K_i) \leq 3066, \quad \lambda_m(K_d) \geq 65 \tag{4.57}$$

In order to evaluate the performance, these controllers are implemented to control the vibration on the excited lab prototype. The control performance is evaluated in terms of their ability to reduce the relative displacement of each floor of the building.

The proportional, derivative, and integral gains are further adjusted to obtain a higher attenuation. Finally, the PID controller gains are chosen to be

$$k_p = 635, \quad k_i = 3000, \quad k_d = 65 \tag{4.58}$$

and the PD controller gains are

$$k_p = 635, \quad k_d = 65 \tag{4.59}$$

Theorem 4.1.1 requires that the PD controller gains need to be positive. In the experiments, the negative gains resulted in an unstable closed-loop operation, which satisfies the conditions in Theorem 4.1.1. Since Theorem 4.2.1 provides sufficient conditions, violating it does not mean instability. We have found that, when $\lambda_M(K_i)$ is more than 4200, the system becomes unstable. This satisfies the condition as (4.57).

Table 4.1 shows the mean squared error, $MSE = \frac{1}{N} \sum_{i=1}^{N} \mathbf{e}_i^2$ of the displacement with proposed controllers, here N is the number of data samples and $\mathbf{e} = (\mathbf{x}^d - \mathbf{x}) = -\mathbf{x}$, where $\mathbf{x}$ is the position achieved using the controllers. Figures 4.9, 4.10, 4.11 and 4.12 show the time response of the first and second floor displacements for both controlled and uncontrolled cases. The control algorithm outputs are shown in Figs. 4.13 and 4.14.

From Table 4.1 one can observe that the controllers effectively decrease the vibration. The controlled response using the PD controller is reduced significantly by applying a damping provided by the derivative gain. Figures 4.10 and 4.12 show the vibration attenuation achieved by adding an integral action to the above PD controller. The results demonstrate that PID controller performs better than PD controller.

Table 4.1 Comparison of vibration attenuation obtained using PD and PID

Control action	PD control	PID control	No control
Floor-1 displacement	0.1699	0.1281	1.0688
Floor-2 displacement	0.5141	0.3386	3.3051

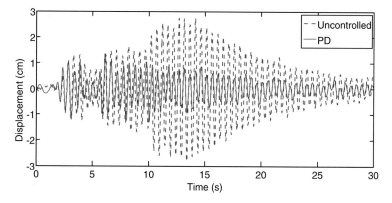

Fig. 4.9 The displacements of the first floor using PD control

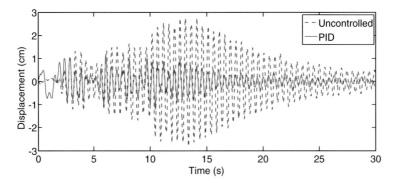

Fig. 4.10 The displacements of the first floor using PID control

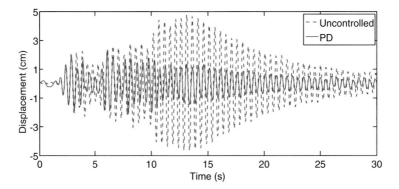

Fig. 4.11 The displacements of the second floor using PD control

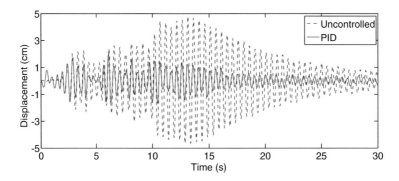

Fig. 4.12 The displacements of the second floor using PID control

Remark 4.3.1 It is worth to note the frequency characteristics of an integrator. An ideal integrator acts like a low-pass filter. The bode magnitude plot of an ideal integrator is shown in Fig. 4.15. At 1.6 Hz the integrator attenuates the input power by 20 dB and at 16 Hz it reaches to 40 dB. During earthquake the structure oscillates

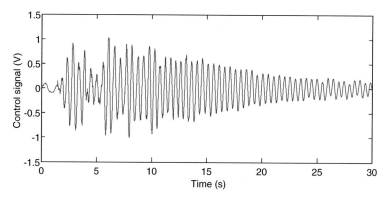

Fig. 4.13 Control signal of PD control

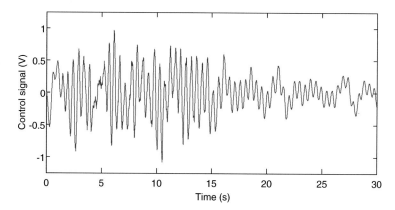

Fig. 4.14 Control signal of PID control

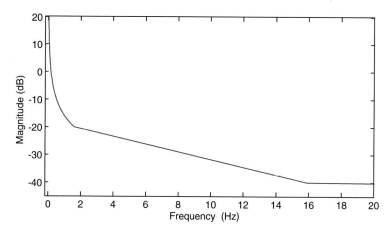

Fig. 4.15 Bode magnitude plot of an ideal integrator

at its natural frequencies. If the natural frequency is very small, then the integrator produces a larger output. The structure prototypes we used for the experiments have natural frequencies 2.1 and 8.9 Hz. Since these frequencies have an attenuation more than 20 dB, a larger value can be used for K_i. On the other hand, if the building has a natural frequency less than 1.6 Hz, then the integral gain should be reduced accordingly. The error input to the integrator is the position data. From Fig. 4.9, 4.10, 4.11, and 4.12, we can see that the position data for the most part takes successive positive and negative values. Hence, the integrator output for high-frequency input signal is small due to the rapid cancellation between these positive and negative values.

Sometimes, the integral control results in an actuator saturation. But as discussed in Remark 2, the output of the integrator is small in our case. Figure 4.16 shows the magnitude spectrum of control signals of the PD and PID controllers. As the building structure is excited mainly in its natural frequency (2.1 Hz), the major control action occurs in this zone. Even though the K_i gain is large, PID controller produces less control effort than the PD controller, but still achieves a better vibration attenuation.

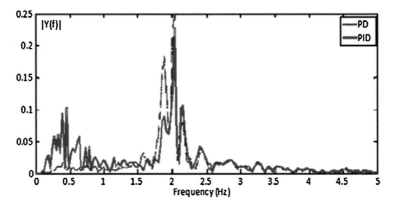

Fig. 4.16 Fourier spectrums of PD and PID control signals

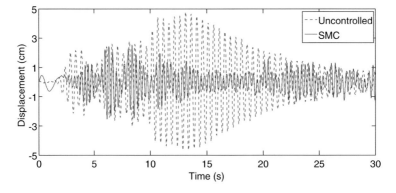

Fig. 4.17 The displacements of the second floor using SMC control

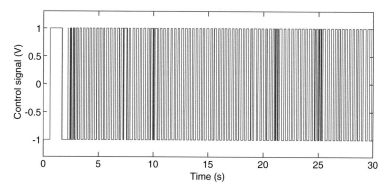

Fig. 4.18 Control signal of SMC control

Remark 4.3.2 From our experience, the classic SMC performs poor, while **x** starts damping from a large to a small value for the shaking table. In Fig. 4.17, after 22 s we can see that the vibration level increases. This is due to the fact that SMC switches aggressively with a gain of η, even though the actual vibration is considerably small, see Fig. 4.18.

References

1. G.W. Housner, L.A. Bergman, T.K. Caughey, A.G. Chassiakos, R.O. Claus, S.F. Masri, R.E. Skelton, T.T. Soong, B.F. Spencer, J.T.P. Yao, Structural control: past, present and future. J. Eng. Mech. **123**, 897–974 (1997)
2. N.R. Fisco, H. Adeli, Smart structures: part I—active and semi-active control. Sci.Iran. **18**, 275–284 (2011)
3. N.R. Fisco, H. Adeli, Smart structures: part II—hybrid control systems and control strategies. Sci. Iran. **18**, 285–295 (2011)
4. M.D. Symans, M.C. Constantinou, Semi-active control systems for seismic protection of structures: a state-of-the-art review. Eng. Struct. **21**, 469–487 (1999)
5. B.F. Spencer, M.K. Sain, Controlling buildings: a new frontier in feedback. IEEE Control Syst. Mag. Emerg. Technol. **17**, 19–35 (1997)
6. T.K. Datta, A state-of-the-art review on active control of structures. ISET J. Earthq. Technol. **40**, 1–17 (2003)
7. J.C.H. Chang, T.T. Soong, Structural control using active tuned mass damper. J. Eng. Mech. ASCE **106**, 1091–1098 (1980)
8. W. Park, K.S. Park, H.M. Koh, Active control of large structures using a bilinear pole-shifting transform with H_∞ control method. Eng. Struct. **30**, 3336–3344 (2008)
9. R. Saragih, Designing active vibration control with minimum order for flexible structures, in *IEEE International Conference on Control and Automation* (2010), pp. 450–453
10. H. Du, N. Zhang, H_∞ control for buildings with time delay in control via linear matrix inequalities and genetic algorithms. Eng. Struct. **30**, 81–92 (2008)
11. K. Seto, A structural control method of the vibration of flexible buildings in response to large earthquake and strong winds, in *Proceedings of the 35th Conference on Decision and Control*, Kobe, Japan (1996)

12. A. Alavinasab, H. Moharrami, Active control of structures using energy-based LQR method. Comput. Aided Civ. Inf. Eng. **21**, 605–611 (2006)
13. J.N. Yang, J.C. Wu, A.K. Agrawal, S.Y. Hsu, Sliding mode control with compensator for wind and seismic response control. Earthq. Eng. Struct. Dyn. **26**, 1137–1156 (1997)
14. J.T. Kim, H.J. Jung, I.W. Lee, Optimal structural control using neural networks. J. Eng. Mech. **126**, 201–205 (2000)
15. D.A. Shook, P.N. Roschke, P.Y. Lin, C.H. Loh, GA-optimized fuzzy logic control of a large-scale building for seismic loads. Eng. Struct. **30**, 436–449 (2008)
16. K.J. Åström, T. Hagglund, Revisiting the Ziegler-Nichols step response method for PID control. J. Process Control **14**, 635–650 (2004)
17. A.C. Nerves, R. Krishnan, Active control strategies for tall civil structures, in *Proceedings of IEEE, International Conference on Industrial Electronics, Control, and Instrumentation*, vol. 2 (1995), pp. 962–967
18. R. Guclu, H. Yazici, Vibration control of a structure with ATMD against earthquake using fuzzy logic controllers. J. Sound Vibr. **318**, 36–49 (2008)
19. R. Guclu, Sliding mode and PID control of a structural system against earthquake. Math. Comput. Model. **44**, 210–217 (2006)
20. T.L. Teng, C.P. Peng, C. Chuang, A study on the application of fuzzy theory to structural active control. Comput. Meth. Appl. Mech. Eng. **189**, 439–448 (2000)
21. S. Thenozhi, W. Yu, Advances in modeling and vibration control of building structures. Ann. Rev. Control **37**(2), 346–364 (2013)
22. A.K. Chopra, *Dynamics of Structures-Theory and Application to Earthquake Engineering*, 2nd edn. (Prentice Hall, Upper Saddle River, 2001)
23. E.D. Sontag, Y. Wang, On characterizations of the input-to-state stability property. Syst. Control Lett. **24**, 351–359 (1995)
24. F.L. Lewis, D.M. Dawson, C.T. Abdallah, *Robot Manipulator Control: Theory and Practice*, 2nd edn. (Marcel Dekker, Inc., New York, 2004)
25. K.J. Åström, T. Hägglund, C.C. Hang, W.K. Ho, Automatic tuning and adaptation for PID controllers-a survey. Control Eng. Pract. **1**, 699–714 (1993)
26. J. Alvarez-Ramirez, R. Kelly, I. Cervantes, Semiglobal stability of saturated linear PID control for robot manipulators. Automatica **39**, 989–995 (2003)
27. X. Yu, Sliding-mode control with soft computing: a survey. IEEE Trans. Ind. Electron. **56**, 3275–3285 (2009)

Chapter 5
Fuzzy PID Control of Building Structures

Abstract In this chapter, the proposed control algorithms combine the classical (PD/PID) and intelligent (Fuzzy) control techniques to achieve good vibration attenuation. The stability analysis of the structural vibration control system is performed using Lyapunov theory and sufficient conditions for tuning the controllers are derived.

Keywords Fuzzy control · PID control · Stability

The objective of structural control is to reduce the vibrations of the buildings due to earthquake or large winds through an external control force. In active control system, it is essential to design an effective control strategy, which is simple, robust, and fault tolerant. Many attempts have been made to introduce advanced controllers for the active vibration control of building structures as discussed in Chap. 2.

PID control is widely used in industrial applications. Without model knowledge, PID control may be the best controller in real-time applications [1]. The great advantages of PID control over the others are that they are simple and have clear physical meanings. Although the research in PID control algorithms is well established, their applications in structural vibration control are still not well developed. In [2], a simple proportional (P) control is applied to reduce the building displacement due to wind excitation. In [3, 4], PD and PID controllers were used. However, the control results are not satisfactory. There are two reasons: (1) it is difficult to tune the PID gains to guarantee good performances such as the rise-time, overshoot, settling time, and steady-state error [3]; (2) in order to decrease the regulation error of PD/PID control, the derivative gain and integration gain have to be increased. These can cause undesired transient performances, even instability [1].

Instead of increasing PD/PID gains, a natural way is to use intelligent method to compensate the regulation error. The difference between our controller and the above intelligent method is that the main part of our controller is still classical PD/PID control. The obstacle of this kind of PD/PID controller is the theoretical difficulty in analyzing its stability. Even for linear PID, it is not easy to prove its asymptotic stability [5].

© The Author(s) 2016
W. Yu and S. Thenozhi, *Active Structural Control with Stable
Fuzzy PID Techniques*, SpringerBriefs in Applied Sciences and Technology,
DOI 10.1007/978-3-319-28025-7_5

In this chapter, the well-known PD/PID is extended to PD/PID control with fuzzy compensation. The stability of these novel fuzzy PD/PID control is proven. Explicit conditions for choosing PID gains are given. Unlike the other PD/PID control for the building structure, the proposed fuzzy PD/PID control does not need large derivative and integral gains. An active vibration control system for a six-story building structure equipped with an AMD is constructed for the experimental study. The experimental results are compared with the other controllers, and the effectiveness of the proposed algorithms are demonstrated.

5.1 Control of Building Structures

The n-floor structure can be expressed as

$$M\ddot{x}(t) + C\dot{x}(t) + F_s = -F_e \tag{5.1}$$

In a simplified case, the lateral force F_s can be linear with x as $F_s = Kx(t)$. However, in the case of real building structures, the stiffness component is inelastic as discussed in the second chapter. Here we consider the nonlinear stiffness. The main objective of structural control is to reduce the movement of buildings into a comfortable level. In order to attenuate the vibrations caused by the external force, an AMD is installed on the structure, see Fig. 4.1. The closed-loop system with the control force $u \in \Re^n$ is defined as

$$M\ddot{x}(t) + C\dot{x}(t) + F_s + F_e = \Gamma(u - \psi) \tag{5.2}$$

where $\psi \in \Re^n$ is the damping and friction force of the damper and $\Gamma \in \Re^{n \times n}$ is the location matrix of the dampers, defined as follows.

$$\Gamma_{i,j} = \begin{cases} 1 \text{ if } i = j = v \\ 0 \text{ otherwise} \end{cases}, \ \forall i, j \in \{1, \dots, n\}, v \subseteq \{1, \dots, n\}$$

where v are the floors on which the dampers are installed. In the case of a two-story building, if the damper is placed on second floor, $v = \{2\}$, $\Gamma_{2,2} = 1$. If the damper is placed on both first and second floor, then $v = \{1, 2\}$, $\Gamma_{2 \times 2} = I_2$.

The damper force F_{dq}, exerted by the qth damper on the structure is

$$F_{dq} = m_{dq}(\ddot{x}_v + \ddot{x}_{dq}) = u_q - \psi_q \tag{5.3}$$

where m_{dq} is the mass of the qth damper, $\ddot{x}_v$ is the acceleration of the vth floor on which the damper is installed, $\ddot{x}_{dq}$ is the acceleration of the qth damper, u_q is the control signal to the qth damper, and

$$\psi_q = c_{dq}\dot{x}_{dq} + \varkappa_q m_{dq} g \ \tanh\left[\beta_t \dot{x}_{dq}\right] \tag{5.4}$$

where c_{dq} and $\dot{x}_{dq}$ are the damping coefficient and velocity of the qth damper, respectively, and the second term is the Coulomb friction represented using a hyperbolic tangent dependent on β_t where $\varkappa_q$ is the friction coefficient between the qth damper and the floor on which it is attached and g is the gravity constant [6].

Obviously, the building structures in open-loop are asymptotically stable when there is no external force, $F_e = 0$. This is also true in the case of inelastic stiffness, due to its BIBO stability and passivity properties [7]. During external excitation, the ideal active control force required for cancelling out the vibration completely is $\Gamma u = F_e$. However, it is impossible because F_e is not always measurable and is much bigger than any control device force. Hence, the objective of the active control is to maintain the vibration as small as possible by minimizing the relative movement between the structural floors. In the next section, we will discuss several stable control algorithms.

5.2 PD Controller with Fuzzy Compensation

PD control may be the simplest controller for the structural vibration control system, see Fig. 5.1, which provides high robustness with respect to the system uncertainties. PD control has the following form

$$u = -K_p(x - x^d) - K_d(\dot{x} - \dot{x}^d) \tag{5.5}$$

where K_p and K_d are positive definite constant matrices, which correspond to the proportional and derivative gains, respectively, and x^d is the desired position. In active vibration control of building structures, the references are $x^d = \dot{x}^d = 0$, hence (5.5) becomes

$$\Gamma u = -K_p x - K_d \dot{x} \tag{5.6}$$

The aim of the controller design is to choose suitable gains K_p and K_d in (5.6), such that the closed-loop system is stable. Without loss of generality, we use a two-story building structure as shown in Fig. 5.1. The nonlinear dynamics of the structure

Fig. 5.1 PD/PID control for a two-story building

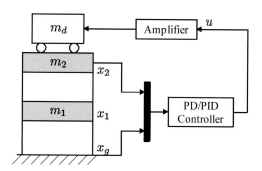

with control can be written as

$$M\ddot{x} + C\dot{x} + F = u \tag{5.7}$$

where

$$F = F_s(x, \dot{x}) + F_e + \psi \tag{5.8}$$

Then the building structure with the PD control (5.6) can be written as

$$M\ddot{x} + C\dot{x} + F = - K_p x - K_d \dot{x} \tag{5.9}$$

The closed-loop system (5.9) is nonlinear and the parameters of M, C, and F are unknown. It is well known that using the PD controller the regulation error can be reduced by increasing the gain K_d. The cost of large K_d is that the transient performance becomes slow. Only when $K_d \rightarrow \infty$, the tracking error converges to zero [8]. Also, it is not a good idea to use a large K_d, if the system comprises high-frequency noise signals.

In this chapter, we use fuzzy compensation to estimate F such that the derivative gain K_d is not so large. A generic fuzzy model, provided by a collection of l fuzzy rules (Mamdani fuzzy model [9]) is used to approximate $\widehat{F}_q$

$$R^i: \text{IF } (x \text{ is } A_{1i}) \text{ and } (\dot{x} \text{ is } A_{2i}) \text{ THEN } \widehat{F}_q \text{ is } B_i^q \tag{5.10}$$

where $\widehat{F}_q$ is the estimation of the uncertain force F.

A total of l fuzzy IF-THEN rules are used to perform the mapping from the input vector z to the output vector $\widehat{F} = \left[\widehat{F}_1 \cdots \widehat{F}_n \right]^T$. Here A_{1i}, A_{2i} and B_i^q are standard fuzzy sets. Using product inference, center-average defuzzification, and a singleton fuzzifier, the output of the fuzzy logic system can be expressed as [10]

$$\widehat{F}_q = \left(\sum_{i=1}^{l} w_{qi} \mu_{A_{1i}} \mu_{A_{2i}} \right) / \left(\sum_{i=1}^{l} \mu_{A_{1i}} \mu_{A_{2i}} \right) = \sum_{i=1}^{l} w_{qi} \sigma_i \tag{5.11}$$

where $\mu_{A_{ji}}$ are the membership functions of the fuzzy sets A_{ji}, which represents the jth rule of the ith input, $i = 1, \ldots, n$ and $j = 1, \ldots, l$. The Gaussian functions are chosen as the membership functions.

$$\mu_{A_{ji}} = \exp \frac{- \left(z_i - \hat{z}_{ji} \right)^2}{\rho_{ji}^2} \tag{5.12}$$

where $\hat{z}$ and ρ is the mean and variance of the Gaussian function, respectively. Weight w_{qi} is the point at which $\mu_{B_i^q} = 1$ and $\sigma_i(x, \dot{x}) = \mu_{A_{1i}} \mu_{A_{2i}} / \sum_{i=1}^{l} \mu_{A_{1i}} \mu_{A_{2i}}$. Equation (5.11) can be expressed in matrix form as

Fig. 5.2 Control scheme for
PD/PID controller with
fuzzy compensator.

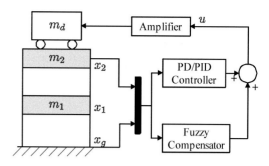

$$\widehat{F} = \widehat{W}\sigma(x,\dot{x}) \tag{5.13}$$

where $\widehat{W} = \begin{bmatrix} w_{11} & \cdots & w_{1l} \\ \vdots & \ddots & \vdots \\ w_{n1} & \cdots & w_{nl} \end{bmatrix}$, $\sigma(x,\dot{x}) = [\sigma_1(x,\dot{x}), \ldots, \sigma_l(x,\dot{x})]^T$.

The PD control with fuzzy compensation, shown in Fig. 5.2 has the following
form.

$$u = -K_p x - K_d \dot{x} - \widehat{W}\sigma(x,\dot{x}) \tag{5.14}$$

In order to analyze the fuzzy PD control (5.14), we define a filtered regulation
error as

$$r = \dot{x} + \Lambda x \tag{5.15}$$

Then the fuzzy PD control (5.14) becomes

$$u = -K_1 r - \widehat{W}\sigma(x,\dot{x}) \tag{5.16}$$

where $K_p = K_1 \Lambda$, $K_d = K_1$, and Λ is a positive definite matrix. Using (5.9), (5.15),
and (5.16):

$$
\begin{aligned}
M\dot{r} &= M(\ddot{x} + \Lambda\dot{x}) \\
&= -C\dot{x} - F - K_1 r - \widehat{W}\sigma(x,\dot{x}) + M\Lambda\dot{x} + C\Lambda x - C\Lambda x \\
&= -K_1 r - \widehat{W}\sigma(x,\dot{x}) - Cr - F + M\Lambda\dot{x} + C\Lambda x \\
&= -K_1 r - \widehat{W}\sigma(x,\dot{x}) - Cr + (M\Lambda\dot{x} + C\Lambda x - F)
\end{aligned}
\tag{5.17}
$$

According to the Universal approximation theorem [10], the general nonlinear
smooth function F can be written as

$$M\Lambda\dot{x} + C\Lambda x - F(x,\dot{x}) = \widehat{W}\sigma(x,\dot{x}) + \phi(x,\dot{x}) \tag{5.18}$$

where $\phi(x, \dot{x})$ is the modeling error which is assumed to be bounded. The following theorem gives the stability analysis of the fuzzy PD control (5.14).

Theorem 5.2.1 *Consider the structural system (5.7) controlled by the fuzzy PD controller (5.16), the closed-loop system is stable, provided that the control gains satisfy*

$$K_1 > 0, \quad K_d > 0 \tag{5.19}$$

The filter regulation error converges to the residual set

$$D_r = \left\{ r \mid \|r\|_{K_1}^2 \leq \bar{\mu}_1 \right\} \tag{5.20}$$

where $\mu^T \Lambda_1^{-1} \mu \leq \bar{\mu}_1$ and $0 < \Lambda_1 < C$.

Proof We define the Lyapunov candidate as

$$V = \frac{1}{2} r^T M r \tag{5.21}$$

Since M and Λ are positive definite matrices, $V \geq 0$. Using (5.17) and (5.18), the derivative of (5.21) is

$$\begin{aligned} \dot{V} &= r^T M \dot{r} \\ &= r^T \left[-K_1 r - \widehat{W} \sigma(x, \dot{x}) - Cr + (M \Lambda \dot{x} + C \Lambda x - F) \right] \\ &= -r^T (K_1 + C) r + r^T \mu \end{aligned} \tag{5.22}$$

The matrix inequality: $X^T Y + Y^T X \leq X^T \Lambda X + Y^T \Lambda^{-1} Y$, is valid for any X, $Y \in \mathfrak{R}^{n \times m}$ and any $0 < \Lambda = \Lambda^T \in \mathfrak{R}^{n \times n}$. Now μ can be estimated as

$$r^T \mu \leq r^T \Lambda_1 r + \mu^T \Lambda_1^{-1} \mu$$

where Λ_1 is any positive definite matrix and we select Λ_1 as

$$C > \Lambda_1 > 0$$

So

$$\dot{V} \leq -r^T (K_1 + C - \Lambda_1) r + \mu^T \Lambda_1^{-1} \mu \tag{5.23}$$

If we choose $K_d > 0$,

$$\dot{V} \leq -r^T K_1 r + \mu \Lambda_1^{-1} \mu = -\|r\|_{K_1}^2 + \bar{\mu}_1 \tag{5.24}$$

where $K_1 > 0$. V is therefore an ISS-Lyapunov function. Using Theorem 1 from [11], the boundedness of $\mu \Lambda_1^{-1} \mu \leq \bar{\mu}_1$ implies that the filter regulation error $r = \dot{x} + \Lambda x$ is bounded, hence x is bounded. Integrating (5.24) from 0 up to T yields

$$V_T - V_0 \leq -\int_0^T r^T K_1 r \mathrm{d}t + \bar{\mu}_1 T \tag{5.25}$$

So

$$\int_0^T r^T K_1 r \mathrm{d}t \leq V_0 - V_T + \bar{\mu}_1 T \leq V_0 + \bar{\mu}_1 T$$

$$\lim_{T \to \infty} \frac{1}{T} \int_0^T \|r\|_{K_1}^2 \, \mathrm{d}t = \bar{\mu}_1 \tag{5.26}$$

$\square$

The approximation accuracy of the fuzzy model (5.13) depends on how to design the membership functions $\mu_{A_{1i}}$, $\mu_{A_{2i}}$, and w_{qi}. In the absence of prior experience, some online learning algorithms can be used to obtain these.

If the premise membership functions A_{1i} and A_{2i} are given by prior knowledge, then $\sigma_i(x, \dot{x}) = \mu_{A_{1i}} \mu_{A_{2i}} / \sum_{i=1}^l \mu_{A_{1i}} \mu_{A_{2i}}$ is known. The objective of the fuzzy modeling is to find the center values of B_{qi} such that the regulation error r is minimized. The fuzzy PD control with automatic updating is

$$\Gamma u = -K_1 r - \widehat{W}_t \sigma(x, \dot{x}) \tag{5.27}$$

The following theorem gives a stable gradient descent algorithm for $\widehat{W}_t$.

Theorem 5.2.2 *If the updating law for the membership function in (5.27) is*

$$\frac{d}{dt}\widehat{W}_t = -K_w \sigma(x, \dot{x}) r^T \tag{5.28}$$

where K_w is a positive definite matrix and

$$K_1 > 0, \quad K_d > 0 \tag{5.29}$$

then the PD control law with fuzzy compensation in (5.14) can make the regulation error stable. In fact, the average regulation error r converges to

$$\limsup_{T \to \infty} \frac{1}{T} \int_0^T \|r\|_{Q_1}^2 \, \mathrm{d}t \leq \bar{\mu}_2 \tag{5.30}$$

where $Q_1 = K_1 + C - \Lambda_2$, $0 < \Lambda_2 < C$, and $\mu^T \Lambda_2^{-1} \mu \leq \bar{\mu}_2$.

Proof The Lyapunov function is

$$V = \frac{1}{2}r^T M r + \frac{1}{2} tr \left(\widetilde{W}_t^T K_w^{-1} \widetilde{W}_t \right) \tag{5.31}$$

where $\widetilde{W}_t = \widehat{W}_t - \widehat{W}$, $\frac{d}{dt}\widetilde{W}_t = \frac{d}{dt}\widehat{W}_t$. Its derivative is

$$\dot{V} = r^T M \dot{r} + tr \left(\widetilde{W}_t^T K_w^{-1} \frac{d}{dt} \widetilde{W}_t \right)$$

$$= r^T \left[-K_1 r - \widehat{W}_t \sigma(x, \dot{x}) - Cr + (M\Lambda\dot{x} + C\Lambda x - F) \right] + tr \left(\widetilde{W}_t^T K_w^{-1} \frac{d}{dt} \widetilde{W}_t \right)$$

$$= -r^T (K_1 + C) r + r^T \mu + r^T \widetilde{W}_t \sigma(x, \dot{x}) + tr \left(\widetilde{W}_t^T K_w^{-1} \frac{d}{dt} \widetilde{W}_t \right)$$

$$= -r^T (K_1 + C) r + r^T \mu + tr \left[\widetilde{W}_t^T \left(K_w^{-1} \frac{d}{dt} \widetilde{W}_t + \sigma(x, \dot{x}) r^T \right) \right] \tag{5.32}$$

If the updating law is (5.28)

$$\dot{V} = -r^T (K_1 + C) r + r^T \mu \tag{5.33}$$

The rest part is similar with the Proof of Theorem 1.   □

Compared with the fuzzy compensation (5.14), the advantage of adaptive fuzzy compensation (5.27) is that, we do not need to be concerned about the big compensation error $\phi(x, \dot{x})$ in (5.18), which results from a poor membership function selection. The gradient algorithm (5.28) ensures that the membership functions $\widehat{W}_t$ is updated such that the regulation error $r(t)$ is reduced. The above theorem also guarantees the updating algorithm is stable.

When we consider the building structure as a black-box, neither the premise nor the consequent parameters are known. Now the objective of the fuzzy compensation is to find $\widehat{W}_t$, as well as the membership functions A_{1i} and A_{2i}. Equation (5.18) becomes

$$\widehat{W}\sigma(x, \dot{x}) - [M\Lambda\dot{x} + C\Lambda x - F(x, \dot{x})]$$

$$= \sum_{i=1}^{l} \left[w_{qi}(t) - \widehat{w}_{qi} \right] z_i^q / b_q + \sum_{i=1}^{l} \sum_{j=1}^{n} \frac{\partial}{\partial \hat{z}_{ji}^q} \left(\frac{a_q}{b_q} \right) \left[\hat{z}_{ji}(t) - \hat{z}_{ji} \right]$$

$$+ \sum_{i=1}^{l} \sum_{j=1}^{n} \frac{\partial}{\partial \rho_{ji}} \left(\frac{a_q}{b_q} \right) \left[\rho_{ji}(t) - \rho_{ji} \right]$$

Define

$$a_q = \sum_{k=1}^{l} w_k \sigma_k, \quad b_q = \sum_{k=1}^{l} \sigma_k, \quad q = 1, 2$$

The updating laws for the membership functions are

$$\frac{d}{dt}\widehat{W}_t = -K_w\sigma(x,\dot{x})r^T$$

$$\frac{d}{dt}\hat{z}_{ji}(t) = -2k_c\sigma_i\frac{\widehat{w}_{qi}-z_i}{b_q}\frac{z_j-\hat{z}_{ji}}{\left[\rho_{ji}^q\right]^2}r^T$$

$$\frac{d}{dt}\rho_{ji}(t) = -2k_\rho\sigma_i\frac{\widehat{w}_{qi}-z_i}{b_q}\frac{\left(z_j-\hat{z}_{ji}\right)^2}{\left[\rho_{ji}\right]^3}r^T \qquad (5.34)$$

The proof is similar with the results in [12].

5.3 PID Controller with Fuzzy Compensation

Although fuzzy compensation can decrease the regulation error of PD control, there still exits regulation error, as given in Theorem 1 and Theorem 2. From control viewpoint, this steady-state error can be removed by introducing an integral component to the PD control. The resulting PID control is given by

$$u = -K_px - K_d\dot{x} - K_i\int_0^t x(\tau)\,d\tau \qquad (5.35)$$

where $K_i > 0$ correspond to the integration gain.

In order to analyze the stability of the PID controller, (5.35) is expressed by

$$u = -K_px - K_d\dot{x} - \xi$$
$$\dot{\xi} = K_ix, \quad \xi(0) = 0 \qquad (5.36)$$

Now substituting (5.36) in (5.7), the closed-loop system can be written as

$$M\ddot{x} + C\dot{x} + F = -K_px - K_d\dot{x} - \xi \qquad (5.37)$$

In matrix form, the closed-loop system is

$$\frac{d}{dt}\begin{bmatrix}\xi\\x\\\dot{x}\end{bmatrix} = \begin{bmatrix}K_ix\\\dot{x}\\-M^{-1}\left(C\dot{x}+F+K_px+K_d\dot{x}+\xi\right)\end{bmatrix} \qquad (5.38)$$

The equilibrium of (5.38) is $\left[x^T,\dot{x}^T,\xi^T\right] = [0,0,\xi^*]$. Since at equilibrium point $x = 0$ and $\dot{x} = 0$, the equilibrium is $\left[0,0,F(0,0)^T\right]$. In order to move the equilibrium to origin, we define

$$\tilde{\xi} = \xi - F(0,0)$$

The final closed-loop equation becomes

$$M\ddot{x} + C\dot{x} + F = -K_p x - K_d \dot{x} - \tilde{\xi} + F(0,0)$$
$$\dot{\tilde{\xi}} = K_i x \tag{5.39}$$

In order to analyze the stability of (5.39), we first give the following properties.

P1. The positive definite matrix M satisfies the following condition.

$$0 < \lambda_m(M) \le \|M\| \le \lambda_M(M) \le \bar{m}, \ \bar{m} > 0 \tag{5.40}$$

where $\lambda_m(M)$ and $\lambda_M(M)$ are the minimum and maximum Eigen values of the matrix M, respectively.

P2. F is Lipschitz over $\bar{x}$ and $\bar{y}$

$$\|F(\bar{x}) - F(\bar{y})\| \le k_F \|\bar{x} - \bar{y}\| \tag{5.41}$$

Most of uncertainties are first-order continuous functions. Since F_s, F_e, and ψ are first-order continuous (C^1) and satisfy Lipschitz condition, **P2** can be established using (5.8). Now we calculate the lower bound of $\int F \, dx$.

$$\int_0^t F \, dx = \int_0^t F_s \, dx + \int_0^t F_e \, dx + \int_0^t \psi \, dx \tag{5.42}$$

We define the lower bound of $\int_0^t F_s \, dx$ is $-\bar{F}_s$ and for $\int_0^t \psi \, dx$ is $-\bar{\psi}$. Compared with F_s and ψ, F_e is much bigger in the case of earthquake. We define the lower bound of $\int_0^t F_e \, dx$ is $-\bar{F}_e$. Finally, the lower bound of $\int_0^t F \, dx$ is

$$k_F = -\bar{F}_s - \bar{F}_e - \bar{\psi} \tag{5.43}$$

The following theorem gives the stability analysis of the PID controller (5.36).

Theorem 5.3.1 *Consider the structural system (5.7) controlled by the PID controller (5.36), the closed-loop system (5.39) is asymptotically stable at the equilibrium* $\left[x^T, \dot{x}^T, \tilde{\xi}^T\right]^T = 0$, *provided that the control gains satisfy*

$$\lambda_m(K_p) \ge \tfrac{3}{2}[k_F + \lambda_M(C)]$$
$$\lambda_M(K_i) \le \beta \frac{\lambda_m(K_p)}{\lambda_M(M)} \tag{5.44}$$
$$\lambda_m(K_d) \ge \beta \left[1 + \frac{\lambda_M(C)}{\lambda_M(M)}\right] - \lambda_m(C)$$

where $\beta = \sqrt{\dfrac{\lambda_m(M)\lambda_m(K_p)}{3}}$.

Proof Here the Lyapunov function is defined as

$$V = \frac{1}{2}\dot{x}^T M \dot{x} + \frac{1}{2}x^T K_p x + \frac{\alpha}{2}\tilde{\xi}^T K_i^{-1}\tilde{\xi} + x^T \tilde{\xi} + \alpha x^T M \dot{x} + \frac{\alpha}{2}x^T K_d x + \int_0^t F dx - k_F$$

(5.45)

where k_F is defined in (5.70) such that $V(0) = 0$. In order to show that $V \geq 0$, it is separated into three parts, such that $V = \sum_{i=1}^3 V_i$

$$V_1 = \frac{1}{6}x^T K_p x + \frac{\alpha}{2}x^T K_d x + \int_0^t F dx - k_F \geq 0$$

(5.46)

$$V_2 = \frac{1}{6}x^T K_p x + \frac{\alpha}{2}\tilde{\xi}^T K_i^{-1}\tilde{\xi} + x^T \tilde{\xi}$$
$$\geq \frac{1}{2}\frac{1}{6}\lambda_m(K_p)\|x\|^2 + \frac{\alpha\lambda_m(K_i^{-1})}{2}\left\|\tilde{\xi}\right\|^2 - \|x\|\left\|\tilde{\xi}\right\|$$

(5.47)

When $\alpha \geq \frac{3}{\lambda_m(K_i^{-1})\lambda_m(K_p)}$,

$$V_2 \geq \frac{1}{2}\left(\sqrt{\frac{\lambda_m(K_p)}{3}}\|x\| - \sqrt{\frac{3}{\lambda_m(K_p)}}\|\xi\|\right)^2 \geq 0$$

(5.48)

and

$$V_3 = \frac{1}{6}x^T K_p x + \frac{1}{2}\dot{x}^T M \dot{x} + \alpha x^T M \dot{x}$$
$$\bar{y}^T A\bar{x} \geq \|\bar{y}\|\|A\bar{x}\| \geq \|\bar{y}\|\|A\|\|\bar{x}\| \geq |\lambda_M(A)|\|\bar{y}\|\|\bar{x}\|$$

(5.49)

when $\alpha \leq \frac{\sqrt{\frac{1}{3}\lambda_m(M)\lambda_m(K_p)}}{\lambda_M(M)}$

$$V_3 \geq \frac{1}{2}\left(\frac{1}{3}\lambda_m(K_p)\|x\|^2 + \lambda_m(M)\|\dot{x}\|^2 + 2\alpha\lambda_M(M)\|x\|\|\dot{x}\|\right)$$
$$= \frac{1}{2}\left(\sqrt{\frac{\lambda_m(K_p)}{3}}\|x\| + \sqrt{\lambda_m(M)}\|\dot{x}\|\right)^2 \geq 0$$

(5.50)

If

$$\sqrt{\frac{1}{3}\lambda_m(K_i^{-1})\lambda_m^{\frac{3}{2}}(K_p)\lambda_m^{\frac{1}{2}}(M)} \geq \lambda_M(M)$$

(5.51)

there exists

$$\frac{\sqrt{\frac{1}{3}\lambda_m(M)\lambda_m(K_p)}}{\lambda_M(M)} \geq \alpha \geq \frac{3}{\lambda_m(K_i^{-1})\lambda_m(K_p)} \tag{5.52}$$

The derivative of (5.70) is

$$\begin{aligned}
\dot{V} &= \dot{x}^T M\ddot{x} + \dot{x}^T K_p x + \alpha\dot{\tilde{\xi}}^T K_i^{-1}\tilde{\xi} + \dot{x}^T\tilde{\xi} + x^T\dot{\tilde{\xi}} + \alpha\dot{x}^T M\dot{x} + \alpha x^T M\ddot{x} + \alpha x^T K_d x + \dot{x}^T F \\
&= \dot{x}^T\left[-C\dot{x} - F - K_p x - K_d\dot{x} - \tilde{\xi} + F(0,0)\right] + \dot{x}^T K_p x + \alpha\dot{\tilde{\xi}}^T K_i^{-1}\tilde{\xi} + \dot{x}^T\tilde{\xi} + x^T\dot{\tilde{\xi}} \\
&\quad + \alpha\dot{x}^T M\dot{x} + \alpha x^T\left[-C\dot{x} - F - K_p x - K_d\dot{x} - \tilde{\xi} + F(0,0)\right] + \alpha x^T K_d\dot{x} + \dot{x}^T F \tag{5.53}
\end{aligned}$$

From (5.41)

$$\alpha x^T[F(0,0) - F] \leq \alpha k_F \|x\|^2$$

Again using the inequality: $X^T Y + Y^T X \leq X^T \Lambda X + Y^T \Lambda^{-1} Y$, we can write

$$-\alpha x^T C\dot{x} \leq \alpha\lambda_M(C)\left[x^T x + \dot{x}^T\dot{x}\right]$$

Since $\dot{\tilde{\xi}} = K_i x$, $\dot{\tilde{\xi}}^T K_i^{-1}\tilde{\xi}$ becomes $\alpha x^T\tilde{\xi}$, and $x^T\dot{\tilde{\xi}}$ becomes $x^T K_i x$, then

$$\dot{V} = -\dot{x}^T[C + K_d - \alpha M - \alpha\lambda_M(C)]\dot{x} - x^T[\alpha K_p - K_i - \alpha k_F - \alpha\lambda_M(C)]x \tag{5.54}$$

Using (5.40), (5.54) becomes,

$$\begin{aligned}
\dot{V} \leq &-\dot{x}^T[\lambda_m(C) + \lambda_m(K_d) - \alpha\lambda_M(M) - \alpha\lambda_M(C)]\dot{x} \\
&- x^T[\alpha\lambda_m(K_p) - \lambda_M(K_i) - \alpha k_F - \alpha\lambda_M(C)]x \tag{5.55}
\end{aligned}$$

If $\lambda_m(C) + \lambda_m(K_d) \geq \alpha[\lambda_M(M) + \lambda_M(C)]$ and $\lambda_m(K_p) \geq \frac{1}{\alpha}\lambda_M(K_i) + k_F + \lambda_M(C)$, then $\dot{V} \leq 0$, hence $\|x\|$ decreases. From (5.52), $\frac{\sqrt{\frac{1}{3}\lambda_m(M)\lambda_m(K_p)}}{\lambda_M(M)} \geq \alpha \geq \frac{3}{\lambda_m(K_i^{-1})\lambda_m(K_p)}$ and $\lambda_m(K_i^{-1}) = \frac{1}{\lambda_M(K_i)}$, if

$$\begin{aligned}
&\lambda_m(K_d) \geq \sqrt{\frac{1}{3}\lambda_m(M)\lambda_m(K_p)\left[1 + \frac{\lambda_M(C)}{\lambda_M(M)}\right]} - \lambda_m(C) \\
&\lambda_m(K_p) \geq \frac{3}{2}[k_F + \lambda_M(C)] \tag{5.56}
\end{aligned}$$

then (5.44) is established.

There exists a ball Σ of radius $\kappa > 0$ centered at the origin of the state-space on which $\dot{V} \leq 0$. The origin of the closed-loop equation (5.39) is a stable equilibrium. Since the closed-loop equation is autonomous, we use La Salle's theorem. Define Ω as

$$\Omega = \left\{ \bar{z}\,(t) = \left[x^T, \dot{x}^T, \tilde{\xi}^T \right]^T \in \Re^{3n} : \dot{V} = 0 \right\}$$
$$= \left\{ \tilde{\xi} \in \Re^n : x = 0 \in \Re^n, \dot{x} = 0 \in \Re^n \right\} \tag{5.57}$$

From (5.72), $\dot{V} = 0$ if and only if $x = \dot{x} = 0$. For a solution $\bar{z}\,(t)$ to belong to Ω for all $t \geq 0$, it is necessary and sufficient that $x = \dot{x} = 0$ for all $t \geq 0$. Therefore, it must also hold that $\ddot{x} = 0$ for all $t \geq 0$. We conclude that from the closed-loop system (5.39), if $\bar{z}\,(t) \in \Omega$ for all $t \geq 0$, then

$$F\,(x, \dot{x}) = F\,(0, 0) = \tilde{\xi} + F\,(0, 0)$$
$$\frac{d}{dt}\xi = 0 \tag{5.58}$$

implies that $\tilde{\xi} = 0$ for all $t \geq 0$. So $\bar{z}\,(t) = 0$ is the only initial condition in Ω for which $\bar{z}\,(t) \in \Omega$ for all $t \geq 0$.

Finally, we conclude from all this that the origin of the closed-loop system (5.39) is asymptotically stable. It establishes the stability of the proposed controller, in the sense that the domain of attraction can be arbitrarily enlarged with a suitable choice of the gains. Namely, increasing K_p the basin of attraction will grow.   □

Remark 5.3.1 Since the stiffness element of the building structure has hysteresis property, its output depends on both the instantaneous and the history of the deformation. The deformation before applying the force (loading) and after removing the force (unloading) is not the same, i.e., the position before the earthquake and after the vibration dies out is not the same. In the absence of external force, the SDOF inelastic structure can be represented as

$$m\ddot{x} + c\dot{x} + f_s(x, \dot{x}) = 0 \tag{5.59}$$

The above system can be described in matrix form as

$$\frac{d}{dt}\begin{bmatrix} x \\ \dot{x} \end{bmatrix} = \begin{bmatrix} \dot{x} \\ -\frac{1}{m}\,(c\dot{x} + f_s(x, \dot{x})) \end{bmatrix} \tag{5.60}$$

The equilibrium of (5.60) is $\left[x^T, \dot{x}^T \right] = \left[\frac{(\bar{\alpha}-1)}{\bar{\alpha}}\tilde{\eta}\,f_r, 0 \right]$, hence the equilibrium position of the system is determined by the nonlinear term $\bar{f}_r$. As a result, after the seismic event, the structural system possibly has infinite number of equilibrium positions. On the other hand, if the system represented in (5.59) is controlled using a PID controller, (as indicated in (4.27)), the integral action force the position asymptotically to zero. However, due to the possibility of variable equilibrium points (this corresponds to the term $F\,(0, 0)$), we cannot conclude that the closed-loop system (5.39) is globally stable.

It is well known that, in the absence of the uncertainties and external forces, the PD control (5.6) with any positive gains can guarantee the asymptotically stable closed-loop system. The main objective of the integral action can be regarded to cancel F. In order to decrease the integral gain, an estimated F is applied to the PID control (5.36). The PID control with an approximate force compensation $\widehat{F}$ is

$$u = -K_p x - K_d \dot{x} - \xi + \widehat{F}, \quad \dot{\xi} = K_i x \tag{5.61}$$

The above theorem can also be applied for the PID controller with an approximate F compensation (5.61). The condition for PID gains (5.44) becomes $\lambda_m\left(K_p\right) \geq \frac{3}{2}\left[\tilde{k}_F + \lambda_M\left(C\right)\right]$ and $\lambda_M\left(K_i\right) \leq \frac{3\beta}{2}\frac{\tilde{k}_F + \lambda_M(C)}{\lambda_M(M)}$, where $\tilde{k}_F \ll k_F$.

However, a big integration gain causes unacceptable transient performances and stability problems. Similar to the fuzzy PD control, a fuzzy compensator for PID control may be used. The fuzzy rules have the same form as (5.10), so the PID control with adaptive fuzzy compensation is

$$u = -K_p x - K_d \dot{x} - K_i \int_0^t x(\tau)\,d\tau - \widehat{W}_t \sigma(x, \dot{x}) \tag{5.62}$$

Hence, the closed-loop system becomes

$$M\ddot{x} + C\dot{x} + F = -K_p x - K_d \dot{x} - \xi - \widehat{W}_t \sigma(x, \dot{x}) \tag{5.63}$$

Similar to the PID, the final closed-loop equation can be written as

$$M\ddot{x} + C\dot{x} + \widehat{W}\sigma(x, \dot{x}) + \phi(x, \dot{x}) = -K_p x - K_d \dot{x} - \tilde{\xi} - \widehat{W}_t \sigma(x, \dot{x}) + \phi(0, 0)$$
$$\dot{\tilde{\xi}} = K_i x \tag{5.64}$$

with the equilibrium $[0, 0, \phi(0, 0)]$, where $\tilde{\xi} = \xi - \phi(0, 0)$ is defined to move the equilibrium point to origin.

In order to analyze the stability of (5.64), we use the following property for $\phi(x, \dot{x})$.

P3. ϕ is Lipschitz over $\bar{x}$ and $\bar{y}$.

$$\|\phi(\bar{x}) - \phi(\bar{y})\| \leq k_\phi \|\bar{x} - \bar{y}\| \tag{5.65}$$

Now we calculate the lower bound of $\int \phi\,dx$ as

$$\int_0^t \phi\,dx = \int_0^t F_s\,dx + \int_0^t F_e\,dx + \int_0^t \psi\,dx - \int_0^t \widehat{W}\sigma\,dx \tag{5.66}$$

Since $\sigma(\cdot)$ is a Gaussian function, $\int_0^t \widehat{W}\sigma\,dx = \frac{\widehat{W}}{2}\sqrt{\pi}\,\mathrm{erf}(z(t))$. Then the lower bound of $\int_0^t \phi\,dx$ is

$$k_\phi = k_F - \frac{1}{2}\sqrt{\pi}\,\widehat{W} \tag{5.67}$$

The following theorem gives the stability analysis of the PID control with adaptive fuzzy compensation (5.62).

Theorem 5.3.2 *Consider the structural system (5.7) controlled by the fuzzy PID controller (5.62), the closed-loop system (5.63) is asymptotically stable, i.e., $\lim_{t\to\infty} x\,(t) = 0$, if the initial condition of $\left[x^T, \dot{x}^T, \tilde{\xi}^T\right]^T$ is inside of Ω and provided that the updating law for the fuzzy compensator is*

$$\frac{d}{dt}\widehat{W}_t = -\left[K_w\sigma(x,\dot{x})(\dot{x} + \alpha x)^T\right]^T \tag{5.68}$$

where K_w is a positive definite matrix and $\alpha > 0$ is a designing parameter and the control gains satisfy

$$\begin{aligned}
\lambda_m\left(K_p\right) &\geq \tfrac{3}{2}\left[k_\phi + \lambda_M\left(C\right)\right]\\
\lambda_M\left(K_i\right) &\leq \beta\frac{\lambda_m(K_p)}{\lambda_M(M)}\\
\lambda_m\left(K_d\right) &\geq \beta\left[1 + \frac{\lambda_M(C)}{\lambda_M(M)}\right] - \lambda_m(C)
\end{aligned} \tag{5.69}$$

Proof We define the Lyapunov function as

$$V = \frac{1}{2}\dot{x}^T M\dot{x} + \frac{1}{2}x^T K_p x + \frac{\alpha}{2}\tilde{\xi}^T K_i^{-1}\tilde{\xi} + \frac{\alpha}{2}x^T K_d x$$

$$+ \int_0^t \phi dx - k_\phi + x^T\tilde{\xi} + \alpha x^T M\dot{x} + \frac{1}{2}tr\left(\widetilde{W}_t^T K_w^{-1}\widetilde{W}_t\right) \tag{5.70}$$

In order to show that $V \geq 0$, it is separated into three parts, where the (5.46) is modified as

$$V_1 = \frac{1}{6}x^T K_p x + \frac{\alpha}{2}x^T K_d x + \int_0^t \phi dx - k_\phi + \frac{1}{2}tr\left(\widetilde{W}_t^T K_w^{-1}\widetilde{W}_t\right) \geq 0 \tag{5.71}$$

whereas the V_2 and V_3 remains the same as in (5.47) and (5.49), respectively. The derivative of (5.70) is

$$\dot{V} = \dot{x}^T M\ddot{x} + \dot{x}^T K_p x + \alpha\dot{\tilde{\xi}}^T K_i^{-1}\tilde{\xi} + \dot{x}^T\tilde{\xi} + x^T\dot{\tilde{\xi}} + \dot{x}^T\phi + \alpha\dot{x}^T M\dot{x}$$

$$+ \alpha x^T M\ddot{x} + \alpha\dot{x}^T K_d x + tr\left(\widetilde{W}_t^T K_w^{-1}\frac{d}{dt}\widetilde{W}_t\right)$$

$$= \dot{x}^T\left[-C\dot{x} - \widetilde{W}_t\sigma(x,\dot{x}) - \phi - K_p x - K_d\dot{x} - \tilde{\xi} + \phi(0,0)\right] + \dot{x}^T K_p x + \dot{x}^T\phi$$

$$+ \alpha\dot{\tilde{\xi}}^T K_i^{-1}\tilde{\xi} + \dot{x}^T\tilde{\xi} + x^T\dot{\tilde{\xi}} + \alpha\dot{x}^T M\dot{x} + \alpha\dot{x}^T K_d x + tr\left(\frac{d}{dt}\widetilde{W}_t^T K_w^{-1}\widetilde{W}_t\right)$$

$$+ \alpha x^T \left[-C\dot{x} - \widetilde{W}_t \sigma(x, \dot{x}) - \phi - K_p x - K_d \dot{x} - \tilde{\xi} + \phi(0, 0) \right]$$

$$= \dot{x}^T \left[-C\dot{x} - \phi - K_p x - K_d \dot{x} - \tilde{\xi} + \phi(0, 0) \right] + \dot{x}^T K_p x + \alpha \tilde{\xi}^T K_i^{-1} \tilde{\xi}$$

$$+ \dot{x}^T \tilde{\xi} + x^T \dot{\tilde{\xi}} + \alpha \dot{x}^T M \dot{x} + \alpha x^T \left[-C\dot{x} - \phi - K_p x - K_d \dot{x} - \tilde{\xi} + \phi(0, 0) \right]$$

$$+ \alpha x^T K_d \dot{x} + \dot{x}^T \phi - tr \left(\frac{d}{dt} \widetilde{W}_t^T K_w^{-1} (\dot{x} + \alpha x)^T \sigma(x, \dot{x}) \right) \widetilde{W}_t \qquad (5.72)$$

If the fuzzy is tuned using (5.68) then

$$\dot{V} = \dot{x}^T \left[-C\dot{x} - K_d \dot{x} + \phi(0, 0) \right] + \alpha \dot{\tilde{\xi}}^T K_i^{-1} \tilde{\xi} + x^T \dot{\tilde{\xi}}$$

$$+ \alpha \dot{x}^T M \dot{x} + \alpha x^T \left[-C\dot{x} - K_p x - \tilde{\xi} + \phi(0, 0) - \phi \right] \qquad (5.73)$$

From (5.65)

$$\alpha x^T \left[\phi(0, 0) - \phi \right] \leq \alpha k_\phi \|x\|^2$$

The rest part is similar with Proof of Theorem 4.3.   □

All the above stability proofs consider that $\Gamma_{n \times n} = I_n$. However in real applications, only few dampers will be utilized for the vibration control, which results in an under-actuated system. In this case, the location matrix Γ should be included along with the gain matrices. In this chapter, we consider only one damper which is installed on the second floor of the structure. For example, in the case of PID controller the control signal becomes,

$$\Gamma u = \begin{bmatrix} 0 & 0 \\ 0 & 1 \end{bmatrix} \left\{ - \begin{bmatrix} k_{p1} & 0 \\ 0 & k_{p2} \end{bmatrix} \begin{bmatrix} x_1 \\ x_2 \end{bmatrix} - \begin{bmatrix} k_{i1} & 0 \\ 0 & k_{i2} \end{bmatrix} \begin{bmatrix} \int_0^t x_1 d\tau \\ \int_0^t x_2 d\tau \end{bmatrix} - \begin{bmatrix} k_{d1} & 0 \\ 0 & k_{d2} \end{bmatrix} \begin{bmatrix} \dot{x}_1 \\ \dot{x}_2 \end{bmatrix} \right\}$$

$$(5.74)$$

$$\Gamma u = \begin{bmatrix} 0 \\ -k_{p2} x_2 - k_{i2} \int_0^t x_2 d\tau - k_{d2} \dot{x}_2 \end{bmatrix} \qquad (5.75)$$

where the scalars k_{p2}, k_{i2}, and k_{d2} are the position, integral, and derivative gains, respectively. In this case, (5.44) becomes,

$$k_{p2} \geq \tfrac{3}{2} \left[k_F + \lambda_M(C) \right]$$

$$k_{i2} \leq \bar{\beta} \frac{\min\{k_{p2}\}}{\lambda_M(M)} \qquad (5.76)$$

$$k_{d2} \geq \bar{\beta} \left[1 + \frac{\lambda_M(C)}{\lambda_M(M)} \right] - \lambda_m(C)$$

where $\bar{\beta} = \sqrt{\dfrac{\lambda_m(M) \min\{k_{p2}\}}{3}}$.

5.4 Experimental Results

The experiment setup, and the relative velocity and position estimatimations are the same as Chap. 4. We compare our fuzzy PD/PID control with the standard PD/PID control and fuzzy control [4]. In order to perform a fair comparison, all the controllers except the fuzzy controller use the same proportional and derivative gains, and same integral gains in the case of PID controllers.

Now, we describe the procedure for selecting the gains for a stable operation. The theorems in this chapter give sufficient conditions for the minimal values of the proportional and derivative gains and maximal values of the integral gains. Here the initial task is to select k_F, which is dominated by the external force F_e. In the experiment, the maximum force used to actuate the building prototype is below 300 N. Hence, we choose $k_F = 365$. Applying these values in Theorem 4.3 we get

$$\lambda_m \left(K_p \right) \geq 556, \ \lambda_M \left(K_i \right) \leq 3066, \ \lambda_m \left(K_d \right) \geq 65$$

Remark 5.4.1 The PID tuning methods are different for the system with and without prior knowledge. If the system parameters are unknown, then auto-tuning techniques are employed to choose the gains either online or off-line. These techniques are broadly classified into direct and indirect methods [13]. In direct method, the closed-loop response of the system is observed and the controller gains are tuned directly based on the past experience and heuristic rules. In the case of indirect method, the structure parameters are identified first from the measured output, and based on these identified parameters the controller is then tuned to achieve the desired system dynamics. This chapter provides a tuning method that ensures a stable closed-loop performance. For that purpose, the structural parameters $\lambda_M (M)$, $\lambda_m(C)$, $\lambda_M(C)$, and k_F, are determined from the identified parameters.

The membership functions of the fuzzy controller in [4] are triangle functions. The position and velocity inputs to this fuzzy system are normalized, such that $x, \dot{x} \in (-1, 1)$. Several experiments showed that nine rules are sufficient to achieve a minimal regulation error.

In our fuzzy PID control, since we use adaptive law, the membership functions are Gaussian functions. Each floor position or velocity is converted into linguistic variables using three membership functions, hence $\widehat{W}^T$, $\sigma \in \mathfrak{R}^9$. We only use the position and velocity of the second floor, and one damper for the control operation, so $r, \widehat{F} \in \mathfrak{R}$. The position and velocity inputs to the adaptive fuzzy system are normalized, such that $x, \dot{x} \in (-1, 1)$. The adaptation rules (5.28) and (5.68) are chosen to be identical by selecting $\Lambda = \alpha$. From (5.52) we choose $\alpha = 6$.

In order to evaluate the performance, these controllers were implemented to control the vibration on the excited lab prototype. The control performance is evaluated in terms of their ability to reduce the relative displacement of each floor of the building.

Table 5.1 Comparison of vibration attenuation obtained using different controllers

Controller	PD	PID	Fuzzy	PD+F	PD+F	No control
Floor-1 (x_1)	0.1669	0.1281	0.0589	0.0255	0.0246	1.0688
Floor-2 (x_2)	0.5141	0.3386	0.1434	0.0733	0.0615	3.3051
Control (u)	0.1232	0.0993	0.1124	0.1408	0.1320	0.0000

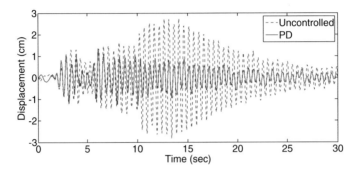

Fig. 5.3 Uncontrolled and controlled displacements of the first floor using PD controller

The proportional, derivative, and integral gains are further adjusted to obtain a higher attenuation. Finally, the PID controller gains are chosen as

$$k_p = 635, \ k_i = 3000, \ k_d = 65$$

and the PD controller gains are

$$k_p = 635, k_d = 65$$

Table 5.1 shows the mean squared error, $MSE = \frac{1}{N}\sum_{i=1}^{N} e_i^2$ of the displacement with proposed controllers, here N is the number of data samples and $e = (x^d - x) = -x$, where x is the position achieved using the controllers. The last row of the Table 5.1 gives the MSE of control signals of each controller $\left(\frac{1}{N}\sum_{i=1}^{N} u_i^2\right)$ with respect to the no control case. Figures 5.3, 5.4, 5.5, 5.6, 5.7, 5.8, 5.9, 5.10, 5.11 and 5.12 show the time response of the first and second floor displacements for both controlled and uncontrolled cases. The control algorithm outputs are shown in Figs. 5.13, 5.14, 5.15, 5.16 and 5.17.

From Table 5.1, one can observe that the controllers effectively decrease the vibration. The controlled response using the PD controller is reduced significantly by providing a damping using the derivative gain. Figures 5.3 and 5.4 show the vibration attenuation achieved by adding an integral action to the above PD controller. The results demonstrate that the PID controller performs better than the PD controller. From Figs. 5.5 and 5.6, it can be seen that the Fuzzy controller achieves more attenu-

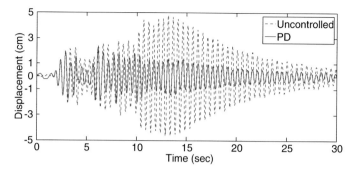

Fig. 5.4 Uncontrolled and controlled displacements of the second floor using PD controller

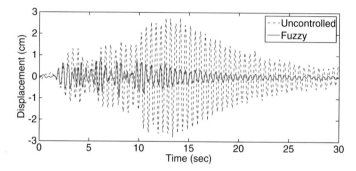

Fig. 5.5 Uncontrolled and controlled displacements of the first floor using fuzzy controller

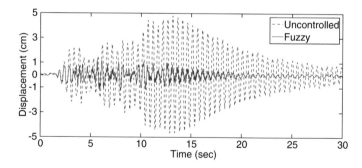

Fig. 5.6 Uncontrolled and controlled displacements of the second floor using fuzzy controller

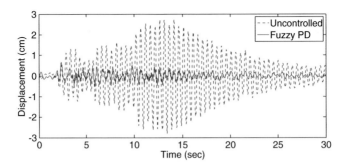

Fig. 5.7 Uncontrolled and controlled displacements of the first floor using fuzzy PD controller

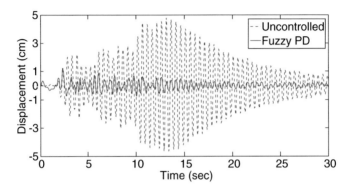

Fig. 5.8 Uncontrolled and controlled displacements of the second floor using fuzzy PD controller

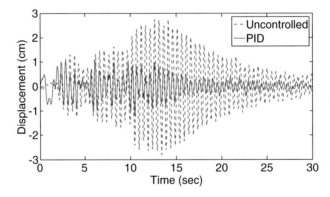

Fig. 5.9 Uncontrolled and controlled displacements of the first floor using PID controller

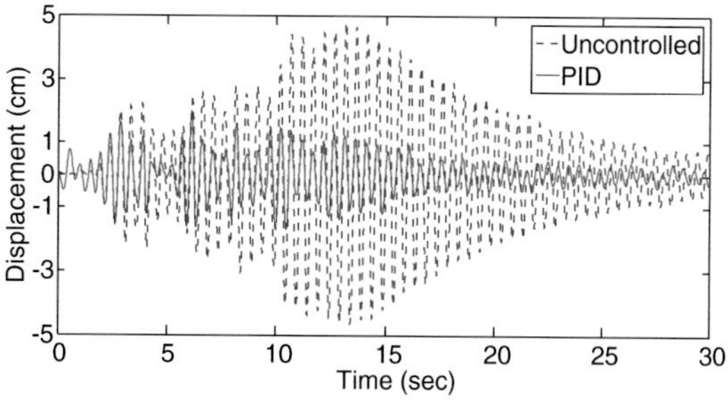

Fig. 5.10 Uncontrolled and controlled displacements of the second floor using PID controller

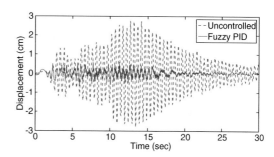

Fig. 5.11 Uncontrolled and controlled displacements of the first floor using fuzzy PID controller

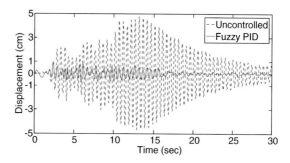

Fig. 5.12 Uncontrolled and controlled displacements of the second floor using fuzzy PID controller

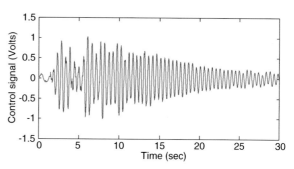

Fig. 5.13 Control signal from PD controller

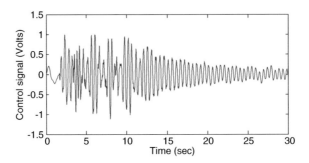

Fig. 5.14 Control signal from fuzzy controller

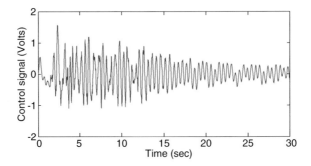

Fig. 5.15 Control signal from fuzzy PD controller

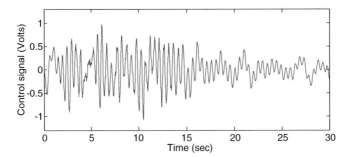

Fig. 5.16 Control signal from PID controller

ation compared to the PD/PID control. Figures 5.7, 5.8, 5.11, and 5.12 illustrate that the structure response reduction can be maximized by the addition of fuzzy compensation to the standard PD/PID control. The performance improvement is based on the fact that this adaptive fuzzy algorithm estimates the control force and also compensate the nonlinear and uncertain forces. From Table 5.1 we can conclude that the fuzzy PID achieves the maximum attenuation.

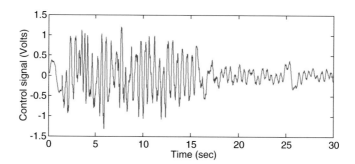

Fig. 5.17 Control signal from fuzzy PID controller

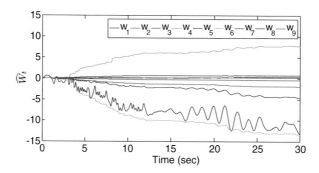

Fig. 5.18 Adaptation of fuzzy weights in fuzzy PD control

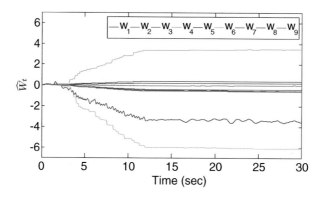

Fig. 5.19 Adaptation of fuzzy weights in fuzzy PID control

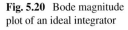

Fig. 5.20 Bode magnitude
plot of an ideal integrator

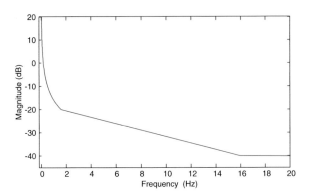

The fuzzy weights adaptation of the fuzzy PD and fuzzy PID control are shown
in Figs. 5.18 and 5.19, respectively. In structural vibration control case, the weight
matrix does not necessarily converge to a constant value. If there exists a position
or a velocity error due to an excitation, then the weights keep changing, see (5.28)
and (5.68). The adaptation of the fuzzy weights of the fuzzy PID control is less than
fuzzy PD, due to the integral action.

Remark 5.4.2 It is worth to note the frequency characteristics of an integrator. An
ideal integrator acts like a low-pass filter. The bode magnitude plot of an ideal integrator is shown in Fig. 5.20. At 1.6 Hz the integrator attenuates the input power by
20 dB and at 16 Hz it reaches to 40 dB. During earthquakes, the structure oscillates
at its natural frequencies. If the natural frequency is very small then the integrator
produces a larger output. The structure prototype we used for the experiments has
natural frequencies 2.1 and 8.9 Hz. Since these frequencies have an attenuation more
than 20 dB a larger value can be used for K_i. On the other hand, if the building
has a natural frequency less than 1.6 Hz, then the integral gain should be reduced
accordingly. The error input to the integrator is the position data. From Figs. 5.3, 5.4,
5.5, 5.6, 5.7, 5.8, 5.9, 5.10, 5.11 and 5.12 we can see that the position data for the
most part takes successive positive and negative values. Hence, the integrator output
for high-frequency input signal is small due to the rapid cancellation between these
positive and negative values.

Figure 5.21 shows the magnitude spectrum of control signals of the simple PID
and fuzzy PID controllers. As the building structure is excited mainly in its natural
frequency (2.1 Hz), the major control action occurs in this zone. In this region the
fuzzy PID controller produces less control effort than the normal PID controller, but
still achieves a better vibration attenuation. Also from Table 5.1 one can see that,

Fig. 5.21 Fourier spectrum of control signals

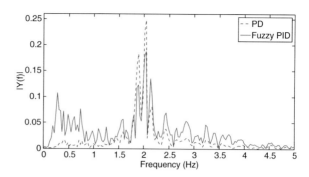

even a small increase in the control action (due to the fuzzy compensation) results in a remarkable vibration attenuation. Sometimes, the integral control results in an actuator saturation. But as discussed in Remark 4.3, the output of the integrator is small in our case.

References

1. K.J. Åström, T. Hagglund, Revisiting the Ziegler-Nichols step response method for PID control. J. Process Control **14**, 635–650 (2004)
2. A.C. Nerves, R. Krishnan, Active control strategies for tall civil structures, in *Proceedings of IEEE, International Conference on Industrial Electronics, Control, and Instrumentation*, vol. 2 (1995), pp. 962–967
3. R. Guclu, Sliding mode and PID control of a structural system against earthquake. Mathe. Comput. Model. **44**, 210–217 (2006)
4. R. Guclu, H. Yazici, Vibration control of a structure with ATMD against earthquake using fuzzy logic controllers. J. Sound Vib. **318**, 36–49 (2008)
5. R. Kelly, A tuning procedure for stable PID control of robot manipulators. Robotica **13**, 141–148 (1995)
6. C. Roldán, F.J. Campa, O. Altuzarra, E. Amezua, *Automatic Identification of the Inertia and Friction of an Electromechanical Actuator*, vol. 17, New Advances in Mechanisms, Transmissions and Applications (Springer, Dordrecht, 2014), pp. 409–416
7. F. Ikhouane, V. Mañosa, J. Rodellar, Dynamic properties of the hysteretic Bouc-Wen model. Syst. Control Lett. **56**, 197–205 (2007)
8. F.L. Lewis, D.M. Dawson, C.T. Abdallah, *Robot Manipulator Control: Theory and Practice*, 2nd edn. (Marcel Dekker Inc., New York, 2004)
9. E.H. Mamdani, Application of fuzzy algorithms for control of simple dynamic plant, *IEE Proceedings*, vol. 121 (1974), pp. 1585–1588
10. L.X. Wang, *Adaptive Fuzzy Systems and Control: Design and Stability Analysis* (PTR Prentice Hall, Upper Saddle River, 1994)
11. E.D. Sontag, Y. Wang, On characterizations of the input-to-state stability property. Syst. Control Lett. **24**, 351–359 (1995)
12. W. Yu, M.A. Moreno-Armendarizb, F.O. Rodrigueza, Stable adaptive compensation with fuzzy CMAC for an overhead crane. Inf. Sci. **181**, 4895–4907 (2011)
13. K.J. Åström, T. Hägglund, C.C. Hang, W.K. Ho, Automatic tuning and adaptation for PID controllers-a survey. Control Eng. Pract. **1**, 699–714 (1993)

Chapter 6
Fuzzy Sliding Mode Control for Wind-Induced Vibration

Abstract In this chapter, we used fuzzy logic to approximate the standard sliding surface and designed a dead-zone adaptive law for tuning the switching gain of the sliding mode control. The stability of the proposed controller is established using Lyapunov stability theory.

Keywords Sliding mode control · Fuzzy control

Active vibration control of building structures under wind and earthquake loadings is a popular field among civil and mechanical engineers. Different control devices and algorithms were proposed and implemented in the last few decades [1, 2]. One of the main challenges in the structural control design is the presence of uncertainties in the building structures, especially in parametric level. Robust control is a well-established technique, which can deal with these uncertainties and disturbances present in the real systems like the building structures.

SMC is one of the most popular robust controllers, which is often seen in the structural vibration control applications. A modal space SMC method is proposed in [3], where only the dominant frequency mode is considered in the design. Another SMC based on the modal analysis is presented in [4], which considers the first six modes. A decentralized system with SMC is presented in [5], where the reaching laws were derived, with and without considering actuator saturations.

Although standard SMC is simple and robust, it does have some limitations. Due to the imperfection in the high-frequency discontinuous switching, the direct implementation of the SMC will result in chattering effect, which may cause damage to the mechanical components like the actuators. The switching gain is selected such that it can overcome the system uncertainties and disturbances. However, the proper gain selection is difficult in the presence of system uncertainty. Different modifications were brought into the standard SMC in the last few decades, which overcome many limitations of the SMC. Higher order sliding mode is one of the popular among them, which reduces chattering. However, its design needs proper tuning of its gains, which requires the knowledge of the uncertainty bounds. In [6], different adaptation techniques were discussed, which are broadly classified into the gain adaptation (Sigma adaptation) and equivalent control (Dynamic adaptation)

© The Author(s) 2016
W. Yu and S. Thenozhi, *Active Structural Control with Stable Fuzzy PID Techniques*, SpringerBriefs in Applied Sciences and Technology, DOI 10.1007/978-3-319-28025-7_6

techniques. Since it is difficult to obtain building parameters, implementation of the equivalent control technique is challenging.

Intelligent control techniques like NN and FLC were also used to design SMC [7]. A NN-based SMC for the active control of seismically excited building structures is proposed in [8]. Here the slope of the sliding surface is considered in the design, which moves in a stable region resulting in a moving sliding surface. To achieve a minimum performance index, this controller is optimized using a genetic algorithm (GA) during the training process. It is shown that a high performing controller is achieved using the moving sliding surface. Another SMC based on radial basis function (RBF) NN is reported in [9]. The chattering-free SMC is obtained using a two-layered RBF NN. The relative displacement of each floor is fed as the input to the NN and the switching gain is derived as the output.

Many research works were carried out in designing the SMC using fuzzy logic the so-called FSMC [10–12]. The SMC provides a stable and fast system, whereas the fuzzy logic provides the ability to handle a nonlinear system. The chattering problem is avoided in most of these FSMC systems. A FSMC based on GA is presented in [13], where the GA is used to find the optimal rules and membership functions for the fuzzy logic controller. Some other structural control strategies based on the non-chattering SMC were also reported [4, 14–16].

The majority of the structural vibration control using SMC [3, 10, 14] uses equivalent control technique. The uncertainty in the building parameters will make them difficult during the implementation. In [15], a low-pass filter is used to estimate the equivalent control. But the filter parameters are difficult to tune and can add phase error to the closed-loop system [6]. SMC with gain adaptation has not yet been discussed in structural vibration control applications. Since the building structure response can be measured, gain adaptation will be a promising technique.

In this chapter, the fuzzy logic and gain adaptation technique were combined for an effective attenuation of the wind-induced vibrations in tall buildings. In order to avoid the chattering phenomenon with respect to the unknown building uncertainty bounds, the sliding mode structural control has been modified in two ways: (1) the sliding surface is approximated using a fuzzy system; and (2) the switching gain of SMC is adapted online. These modifications successfully overcome the problems of the other fuzzy/adaptive SMC, such as the necessity of the equivalent control and the knowledge of the upper bounds of the structure uncertainties. Moreover, the adaptation algorithm guarantees that the switching gain is not overestimated. Theoretically, it has been shown that the proposed controller guarantees a bounded system trajectory and the states can be driven to an arbitrarily small neighborhood of the sliding surface. An active vibration control system for a six-story building structure equipped with an AMD has been constructed for the experimental study. The controller performance was also verified under the seismic excitation. The experimental results were compared with the other controller results and the effectiveness of the proposed algorithms has been demonstrated.

6.1 Control of Wind-Induced Vibration of High-Rise Buildings

The wind force acts on a building in the form of an external pressure, see Fig. 6.1a. The frequency range of wind forces is usually lesser than that of the earthquake forces, see Fig. 6.1b. For that reason the high-rise buildings are more affected by the wind forces. If the wind-induced vibration exceeds more than $0.15\,\mathrm{m/s^2}$, humans may feel uncomfortable and the fragile items in the building may get damaged [17]. The main objective of structural control against the high wind forces is to reduce the relative movement of the building floors into a comfortable level.

If the wind-induced vibration exceeds more than $0.15\,\mathrm{m/s^2}$, humans may feel uncomfortable and the fragile items in the building may get damaged. In order to attenuate the vibrations caused by the external wind force, an AMD is installed on the top floor of the building structure, see Fig. 2.6. Depending on the size of the building, the power requirements of these actuators may vary from kilowatts to several megawatts. So it is important to achieve a satisfying vibration attenuation by keeping the energy requirements as minimum as possible. Moreover, a larger input signal can result in a saturation of the actuator.

In order to attenuate the vibrations caused by the external wind force, an AMD has been installed in the structure, see Fig. 2.6. The force exerted by the AMD on the structure is

$$F_d = m_d(\ddot{x}_r + \ddot{x}_d) = u - d \tag{6.1}$$

where m_d is the mass of the damper, $\ddot{x}_r$ is the acceleration of rth floor on which the damper is installed, $\ddot{x}_d$ is the acceleration of the damper, u is the control input, and

$$d = c_d\dot{x}_d + \epsilon m_d g \operatorname{sign}[\dot{x}_d]$$

where c_d and $\dot{x}_d$ are the damping coefficient and velocity of the damper, respectively, g is the acceleration due to gravity, and ϵ is the friction coefficient between the damper and the floor on which it is attached.

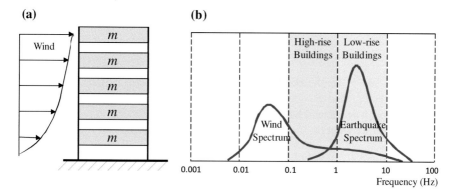

Fig. 6.1 **a** Wind excitation **b** Frequency spectrum of excitations

The closed-loop structural system with AMD is

$$M\ddot{x}(t) + C\dot{x}(t) + F_s(x) + F_e = \Gamma u \tag{6.2}$$

where $u \in \mathfrak{R}^n$ and $\Gamma \in \mathfrak{R}^{n \times n}$ is the location matrix of the dampers, defined as follows:

$$\Gamma_{i,j} = \begin{cases} 1 & \text{if } i = j = r \\ 0 & \text{otherwise} \end{cases}, \quad \forall i, j \in \{1, ..., n\}, r \subseteq \{1, ..., n\}$$

where r indicates the floors on which the dampers are installed. In the case of a two-story building, if the damper is placed on the second floor, $r = \{2\}$, $\Gamma_{2,2} = 1$. If the damper is placed on both the first and second floors, then $r = \{1, 2\}$, $\Gamma_{2 \times 2} = I_2$.

Obviously, the building structures are stable when there is no external force, $F_e = 0$. During excitation, the ideal active control is $\Gamma u = F_e$. However, it is impossible because F_e is not always measurable and $F_e \gg F_d$. Depending on the size of the building, the power requirements of these actuators vary from kilowatts to several megawatts. So it is important to achieve a satisfying vibration attenuation by keeping the energy requirements as minimum as possible. Moreover, a larger control input signal can result in a saturation of the actuator.

Here the structure stiffness of each floor is coupled with its upper floor stiffness. This model assumes that the mass of the structure is concentrated at each floor and the wind force acts horizontally on the structure. The structural parameters such as the mass, damping, and stiffness can be modeled by identifying the building parameters via the dynamic response from its input and output data, see [21].

In the case of real building structures, if the relationship between the lateral force F_s and x is nonlinear, then the stiffness component is inelastic. This occurs when the structure is excited by a very strong force, which deforms the structure beyond its limit of linear elastic behavior. For a single-degree-of-freedom case, the nonlinear force $f_s \in \mathfrak{R}$ can be represented using Bouc–Wen model [18] as

$$f_s(x) = \tilde{\alpha}kx + (1 - \tilde{\alpha})k\tilde{\eta}\varphi \tag{6.3}$$

where $\tilde{\alpha}$, k, and $\tilde{\eta}$ are positive numbers and φ is the nonlinear restoring force that satisfies

$$\dot{\varphi} = \tilde{\eta}^{-1} \left(\tilde{\delta}\dot{x} - \tilde{\beta}|\dot{x}||\varphi|^{p-1}\varphi + \tilde{\gamma}\dot{x}|\varphi|^p \right) \tag{6.4}$$

where $\tilde{\delta}$, $\tilde{\beta}$, and $\tilde{\gamma}$ are the positive numbers and p is an odd number. A physical hysteresis element described using (6.3) and (6.4) is bounded-input bounded-output (BIBO) stable, passive, and has a unique solution [19].

The structure model (6.2) can be rewritten in state-space form as

$$\begin{aligned} \dot{z}_1 &= z_2 \\ \dot{z}_2 &= f(z) + \tilde{\Gamma}u \end{aligned} \tag{6.5}$$

where $z_1 = x$, $z_2 = \dot{x}$, $f(z) = -M^{-1}[Cz_2 + F_s(x, \dot{x}) + F_e]$, $\tilde{\Gamma} = M^{-1}\Gamma$. The output can be defined as $y = Hz$, where H is a known matrix.

One of the most effective approaches for dealing the model uncertainty is the robust control. Equation (6.5) can be written as

$$\dot{z} = \tilde{A}z + \tilde{B}f_0(z) + \tilde{B}\Delta f + \tilde{B}\tilde{\Gamma}u \qquad (6.6)$$

where f_0 is the nominal structure dynamics, Δf is the uncertainty part, $\tilde{A} = \begin{bmatrix} 0 & I_n \\ 0 & 0 \end{bmatrix} \in \mathfrak{R}^{2n \times 2n}$, $\tilde{B} = [0, I_n]^T \in \mathfrak{R}^{2n \times n}$, $z = \left[z_1^T, z_2^T\right]^T \in \mathfrak{R}^{2n}$. We assume that the uncertainty, $\Delta f = f(z) - f_0(z)$ is bounded as

$$\|\Delta f\| \le \bar{f}_d \qquad (6.7)$$

If the parameters of $f(z)$ are completely unknown, then we assume that $f(z)$ is also bounded.

$$\|f(z)\| \le \bar{f} \qquad (6.8)$$

This assumption is practically reasonable, because in the absence of external forces the building structure is stable and the big external excitation forces are also bounded, $\|F_e\| \le \bar{F}_e$.

6.2 Sliding Mode Control with Fuzzy Sliding Surface

In recent years, increasing attention has been given to the systems with discontinuous control actions. By intelligent selection of control actions, the state trajectories of the system under control could be modified correspondingly to give the desired properties. The control design problem in such systems with discontinuous control actions (SMC) can be reduced to the convergence problem to a special surface in the corresponding phase space (sliding surface).

A general class of discontinuous control is defined by the following relationships:

$$u = -\eta P^{-1}\text{sign}(s) = \begin{cases} -\eta P^{-1} & \text{if } s > 0 \\ 0 & \text{if } s = 0 , \\ \eta P^{-1} & \text{if } s < 0 \end{cases} \quad P = P^T > 0 \qquad (6.9)$$

where $\eta > 0$ is the switching gain, s is the sliding surface, and $\text{sign}(s) = \left[\text{sign}(s_1), \ldots, \text{sign}(s_{2n})\right]^T$. The sliding surface can be a function of the regulation error $e = z - z^d$, where z^d is the desired state. If we use $s = e$, then the objective of the SMC is to drive the regulation error to zero in the presence of disturbance. In active vibration control of building structures, the references are defined as $z^d = \left[\left(x^d\right)^T, \left(\dot{x}^d\right)^T\right]^T = 0$, then $s = \left[x^T, \dot{x}^T\right]^T \in \mathfrak{R}^{2n}$ and $\dot{s} = \dot{z}$.

Consider the positive definite quadratic forms

$$V_1 = s^T P \Phi^\dagger s, \quad \Phi = \tilde{B} \tilde{\Gamma} \tag{6.10}$$

where $\Phi^\dagger = \left(\Phi^T \Phi\right)^{-1} \Phi$ is the pseudo-inverse matrix of Φ. Finding the time derivative of function (6.10) on the trajectory of system (6.6), we get

$$\dot{V}_1 = z^T \left(\tilde{A}^T P \Phi^\dagger + P \Phi^\dagger \tilde{A}\right) z + 2z^T P \Phi^\dagger \tilde{B} f + 2z^T P \Phi^\dagger \tilde{B} \tilde{\Gamma} u \tag{6.11}$$

Since $\tilde{A}$ is a stable matrix, there exits $Q = Q^T > 0$, such that $\tilde{A}^T P \Phi^\dagger + P \Phi^\dagger \tilde{A} = -Q$. Using the property $\Phi^\dagger = \Phi^{-1}$, (6.8), and (6.9), we can get

$$\dot{V}_1 \leq -\|z\|_Q^2 + 2\bar{f} \left\| P \Phi^\dagger \tilde{B} \right\| \|z\| - 2z^T P \eta P^{-1} \text{sign}(z) \tag{6.12}$$

Now using the property $z^T \text{sign}(z) = \|z\|$ we can write

$$\begin{aligned}\dot{V}_1 &\leq -\|z\|_Q^2 + 2\bar{f} \left\| P \Phi^\dagger \tilde{B} \right\| \|z\| - 2\eta \|z\| \\ &= -\|z\|_Q^2 + 2\|z\| \left(\bar{f} \left\| P \Phi^\dagger \tilde{B} \right\| - \eta\right) \\ &\leq 2\|z\| \left(\bar{f} \left\| P \Phi^\dagger \tilde{B} \right\| - \eta\right)\end{aligned} \tag{6.13}$$

Obviously, if the gain of the sliding mode control satisfies the following condition

$$\eta \geq \bar{f} \left\| P \Phi^\dagger \tilde{B} \right\|$$

then $\dot{V}_1 \leq 0$. From [20] we know that $s = e$ will converge to zero.

Generally, the civil engineers design the structural parameters such that it can withstand a given load [21]. From this design, one can have an approximation about the upper bound of the structural uncertainty. In the case of robust control such as the classic SMC, the gain is selected to assure a robust performance by considering the worst situation. Hence, by choosing a sufficiently high gain η in (6.9), the effect of any parameter variations can be made negligible. However, this may amplify the chattering effect, where the control signal switches in a high frequency within a tight neighborhood of the sliding surface. In structural control, this is also caused by the unmodeled parasitic dynamics present in the system. This high-frequency switching can damage mechanical systems like the actuators. Although the huge AMD with big time constant in the structural vibration control can be regarded as a second-order low-pass filter and does not respond to high-frequency commands, the chattering control signal may damage the damper's motor mechanism.

Many strategies were proposed to reduce the chattering phenomenon. The boundary layer method approximates the sign function in (6.9) using a saturation function.

$$u = -\eta \text{sat}(s) = \begin{cases} -\eta & \text{if } s > \delta \\ \frac{s}{\delta}\eta & \text{if } \delta \geq s \geq -\delta \\ \eta & \text{if } s < -\delta \end{cases}, \quad \eta > 0 \qquad (6.14)$$

where δ is a positive constant and 2δ is the thickness of the boundary layer. In general, the bigger the boundary layer thickness, the smoother the control signal, and the bigger the residual set to which s will converge. The boundary layer method smooths the control signal with a loss of control accuracy.

In this chapter, we use a fuzzy system to smooth the sliding surface s. We use the following three fuzzy rules:

R^1: IF s is "Positive" P THEN u is "Negative" $-\eta$
R^2: IF s is "Zero" Z THEN u is "Zero" Z
R^3: IF s is "Negative" N THEN u is "Positive" η

The choice of membership functions decides how well a fuzzy system approximates a function. Here the goal is to select the membership functions such that it can approximate the sign function with a smooth switching near the zero vicinity, see Fig. 6.2. The membership function of the input linguistic variable s is defined as μ_A and the membership function of the output linguistic variable u is defined as μ_B. It is straightforward to verify that the membership functions shown in Fig. 6.3 can produce an output surface similar to Fig. 6.2.

Using product inference, center-average, and singleton fuzzifier, the output of the fuzzy logic system can be expressed as

$$u = \eta \frac{w_1 \mu_{A_P}(s) + w_2 \mu_{A_Z}(s) + w_3 \mu_{A_N}(s)}{\mu_{A_P}(s) + \mu_{A_Z}(s) + \mu_{A_N}(s)} \qquad (6.15)$$

where μ_{A_P}, μ_{A_Z}, and μ_{A_N} are the membership functions of "Positive", "Zero", and "Negative" of the input s, w_i, $i = 1, 2, 3$, are the points at which $\mu_B = 1$. From Fig. 6.3 (b) $w_1 = -1$, $w_2 = 0$, $w_3 = 1$, then (6.15) becomes

Fig. 6.2 FSMC switching

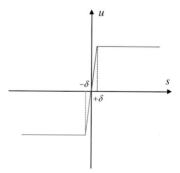

Fig. 6.3 Membership
functions: **a** input set
b output set

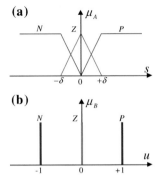

$$u = \frac{\mu_{A_N}(s) - \mu_{A_P}(s)}{\mu_{A_P}(s) + \mu_{A_Z}(s) + \mu_{A_N}(s)} \tag{6.16}$$

We can see that, when $s > \delta$, $\mu_{A_P}(s) = 1$, $\mu_{A_Z}(s) = 0$, $\mu_{A_N}(s) = 0$, then $u = -\eta$; when $s < -\delta$, $\mu_{A_P}(s) = 0$, $\mu_{A_Z}(s) = 0$, $\mu_{A_N}(s) = 1$, then $u = \eta$. Finally, the sliding mode control with fuzzy sliding surface is

$$u = \begin{cases} -\eta P^{-1} \operatorname{sign}(s) & \text{if } \|s\| > \delta \\ \eta \frac{\mu_{A_N}(s) - \mu_{A_P}(s)}{\mu_{A_P}(s) + \mu_{A_Z}(s) + \mu_{A_N}(s)} & \text{if } \|s\| \le \delta \end{cases}, \quad \eta > 0 \tag{6.17}$$

The stability of the fuzzy sliding mode control (6.17) is proved by the same Lyapunov function (6.10). By substituting the FSMC control (6.17) into (6.11), the stability can be concluded using the following two cases:

(1) When $\|s\| > \delta$, $u = -\eta P^{-1}\operatorname{sign}(s)$. It is the same as (6.13), if $\eta \ge \bar{f} \left\| P \Phi^{\dagger} \tilde{B} \right\|$, then $\dot{V}_1 \le 0$, hence s decreases.

(2) When $\|s\| \le \delta$, $u = \eta \frac{\mu_{A_N}(s) - \mu_{A_P}(s)}{\mu_{A_P}(s) + \mu_{A_Z}(s) + \mu_{A_N}(s)}$. Then s is bounded in the residual set δ.

From (1) and (2), we know that s is bounded and the total time during which $\|s\| > \delta$ is finite. Let T_j denotes the time interval during which $\|s\| > \delta$. (a) If only finite times s stay outside the circle of radius δ (and then reenter), s will eventually stay inside this circle. (b) If s leave the circle infinite times, the total time s leave the circle is finite,

$$\sum_{j=1}^{\infty} T_j < \infty, \quad \lim_{j \to \infty} T_j = 0 \tag{6.18}$$

So s is bounded via an invariant set argument. Let $s(j)$ denotes the largest tracking error during the T_j interval. Then (6.18) and bounded $s(j)$ imply that

$$\lim_{j \to \infty} [-s(j) + \delta] = 0$$

Fig. 6.4 Concept of real
sliding surface

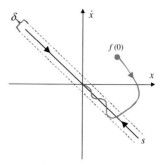

So $s(j)$ will converge to δ. From these discussions, one can say that the implementation of (6.17) can only assure a "real sliding surface" [22], which guarantees that the state trajectories will slide within a domain (δ), see Fig. 6.4.

6.3 Fuzzy Sliding Mode Control with Adaptive Gain

Although the fuzzy sliding mode control (6.17) solves the chattering problem near the sliding surface, it requires a big gain $\eta \geq \bar{f} \, \| P\Phi^\dagger B \|$. This overestimation of the gain is not well advised. In the absence of system boundary knowledge, online gain adaptation can solve this problem. Here the switching gain η in (6.17) is replaced by the adaptive gain η_t, which uses the following adaptive law:

$$\dot{\eta}_t = \begin{cases} \bar{\eta} \, \|s\| \, \text{sign} \, (\|s\| - \delta) & \text{if } \eta_t > \mu \\ 0 & \text{if } \eta_t \leq \mu \end{cases}, \quad \bar{\eta}, \mu > 0 \qquad (6.19)$$

where μ is used to assure that the gain η_t is always positive and δ is the parameter defined in (6.17). When the state trajectory is outside the domain δ, the gain η_t will keep increasing until $\|s\| < \delta$; then the gain starts decreasing, once the state trajectory reaches inside the δ-domain. Thus the overestimation of the switching gain is avoided. Figure 6.5 shows the block diagram of the proposed fuzzy sliding mode control with adaptive gain.

Fig. 6.5 Block diagram of
the AFSMC

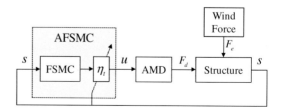

We use the following Lyapunov-like candidate:

$$V_2 = s^T P \Phi^\dagger s + \frac{1}{2\gamma}(\eta_t - \eta^*)^2 \tag{6.20}$$

where $\gamma, \eta^* > 0$. When the gain of the sliding mode is updated by a gradient algorithm as in (6.19), the gain is bounded [23]. We can assume that the upper bound of η_t is η^*.

The time derivative of (6.20) is

$$\dot{V}_2 = z^T \left(\tilde{A}^T P \Phi^\dagger + P \Phi^\dagger \tilde{A} \right) z + 2z^T P \Phi^\dagger \tilde{B} f + 2z^T P \Phi^\dagger \tilde{B} \tilde{\Gamma} u + \frac{1}{\gamma}(\eta_t - \eta^*)\dot{\eta}_t \tag{6.21}$$

Once again using (6.8), (6.9), and $z^T \text{sign}(z) = \|z\|$, we can write (6.21) as

$$\dot{V}_2 \leq - \|z\|_Q^2 + 2\bar{f} \left\| P \Phi^\dagger \tilde{B} \right\| \|z\| - 2z^T P \Phi^\dagger \tilde{B} \tilde{\Gamma} \eta_t P^{-1} \text{sign}(z) + \frac{1}{\gamma}(\eta_t - \eta^*)\dot{\eta}_t$$

$$= - \|z\|_Q^2 + 2 \|z\| \left(\bar{f} \left\| P \Phi^\dagger \tilde{B} \right\| - \eta_t \right) + \frac{1}{\gamma}(\eta_t - \eta^*)\dot{\eta}_t \tag{6.22}$$

Using the adaptive law (6.19), (6.22) becomes

$$\dot{V}_2 \leq - \|z\|_Q^2 + 2 \|z\| \left(\bar{f} \left\| P \Phi^\dagger \tilde{B} \right\| - \eta_t \right) + \frac{1}{\gamma}(\eta_t - \eta^*)\bar{\eta} \|z\| \ \text{sign}(\|z\| - \delta)$$

$$\leq 2 \|z\| \left(\bar{f} \left\| P \Phi^\dagger \tilde{B} \right\| - \eta_t \right) + \frac{1}{\gamma}(\eta_t - \eta^*)\bar{\eta} \|z\| \ \text{sign}(\|z\| - \delta)$$

$$= 2 \|z\| \left(\bar{f} \left\| P \Phi^\dagger \tilde{B} \right\| - \eta^* \right) + 2(\eta_t - \eta^*) \left(- \|z\| + \frac{\bar{\eta}}{2\gamma} \|z\| \ \text{sign}(\|z\| - \delta) \right) \tag{6.23}$$

The above expression can be rewritten as

$$\dot{V}_2 \leq 2 \|z\| \left(\bar{f} \left\| P \Phi^\dagger \tilde{B} \right\| - \eta^* \right) + 2(\eta_t - \eta^*) \left(- \|z\| + \frac{\bar{\eta}}{2\gamma} \|z\| \ \text{sign}(\|z\| - \delta) \right)$$
$$-2 |\eta_t - \eta^*| + 2 |\eta_t - \eta^*| \tag{6.24}$$

Since η^* is the upper bound of η_t, $(\eta_t - \eta^*) = - |\eta_t - \eta^*|$, then (6.24) is

$$\dot{V}_2 \leq -2 \|z\| \left(\eta^* - \bar{f} \left\| P \Phi^\dagger \tilde{B} \right\| \right) - 2 |\eta_t - \eta^*|$$

$$- 2 |\eta_t - \eta^*| \left(- \|z\| + \frac{\bar{\eta}}{2\gamma} \|z\| \ \text{sign}(\|z\| - \delta) - 1 \right) \tag{6.25}$$

Let's define

$$\beta_z = 2 \left\| P \Phi^\dagger \right\|^{\frac{1}{2}} \left(\eta^* - \bar{f} \left\| P \Phi^\dagger \tilde{B} \right\| \right)$$

$$\Psi = 2 |\eta_t - \eta^*| \left(- \|z\| + \frac{\bar{\eta}}{2\gamma} \|z\| \ \text{sign}(\|z\| - \delta) - 1 \right) \tag{6.26}$$

Then (6.25) can be written as

$$\dot{V}_2 \leq - \left\| P \Phi^\dagger \right\|^{\frac{1}{2}} \beta_z \|z\| - 2\left| \eta_t - \eta^* \right| - \Psi$$

$$= - \left\| P \Phi^\dagger \right\|^{\frac{1}{2}} \beta_z \|z\| - \frac{2\sqrt{2\gamma} \, |\eta_t - \eta^*|}{\sqrt{2\gamma}} - \Psi$$

$$\leq - \min \left\{ \beta_z, \sqrt{8\gamma} \right\} \left(\left\| P \Phi^\dagger \right\|^{\frac{1}{2}} \|z\| + \frac{|\eta_t - \eta^*|}{\sqrt{2\gamma}} \right) - \Psi \qquad (6.27)$$

Finally, we have

$$\dot{V}_2 \leq -\beta_v V_2^{\frac{1}{2}} - \Psi \qquad (6.28)$$

where $\beta_v = \min \left\{ \beta_z, \sqrt{8\gamma} \right\}$. The stability of the system depends on the term Ψ. It is obvious that $\dot{V}_2 \leq 0$ when $\Psi = 0$. We consider the following two cases:

(1) $\|s\| = \|z\| > \delta$. If we select γ in (6.20) such that it satisfies

$$0 < \gamma < \frac{\bar{\eta}\delta}{2(\delta + 1)} \qquad (6.29)$$

then $\Psi > 0$, (6.28) is

$$\dot{V}_2 \leq -\beta_v V_2^{\frac{1}{2}} \leq 0 \qquad (6.30)$$

(2) $\|s\| = \|z\| \leq \delta$. From the definition (6.26), $\Psi < 0$. If $|\Psi| > \left| \beta_v V_2^{\frac{1}{2}} \right|$, then $u = \eta_t \frac{\mu_{A_N}(s) - \mu_{A_P}(s)}{\mu_{A_P}(s) + \mu_{A_Z}(s) + \mu_{A_N}(s)}$, the state trajectory may go unstable, i.e., $\|z\|$ will increase until $\|z\| > \delta$, which satisfies the condition (1). Now, $\dot{V}_2 \leq 0$, hence $\|z\|$ will decrease and fall inside the δ-domain. If $|\Psi| \leq \left| \beta_v V_2^{\frac{1}{2}} \right|$, then $\dot{V}_2 \leq 0$ which is same as the condition (1).

When $\|s\| > \delta$, the controller requires time for s to return to the δ-domain. Until then, s will remain in another domain δ_l, where $\delta_l > \delta$. From (6.19) we can see that the rate of change of η_t can be increased by selecting a big $\bar{\eta}$. Increasing the gain $\bar{\eta}$ means that δ_l will decrease toward δ.

6.4   Experimental Results

The experimental setup is similar to that of Chap. 4 with the following changes. Only the linear servo actuator STB1104 is used instead of STB1108, hence the mass (2 % of building mass), and the force constant (5.42 N/A) has been changed.

The objective of the structural control system is to reduce the relative motion between the floors. In the case of wind excitation, the ground acceleration in (2.1), $x_g = 0$. The proposed controller needs the structure position and velocity data. Two accelerometers (Summit Instruments 13203B) are used to measure the ground and

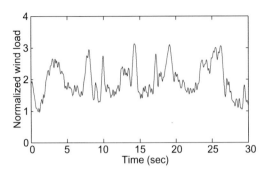

Fig. 6.6 Wind excitation signal

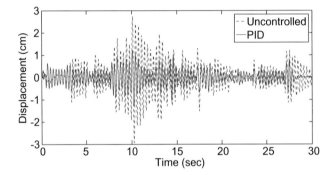

Fig. 6.7 Uncontrolled and controlled displacements of the top floor using PID controller

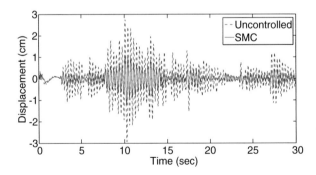

Fig. 6.8 Uncontrolled and controlled displacements of the top floor using SMC

the top floor accelerations. The ground acceleration is then subtracted from the top floor acceleration to get the relative floor movement. The velocity and position data are then estimated using the numerical integrator proposed in Chap. 3.

The proposed AFSMC performance is compared with the classic PID controller and normal SMC. All of these controllers are designed to work within the normal operation range of the AMD. The PID control given in (5.35) has been used here, with gains $k_p = 425$, $k_i = 50$, $k_d = 55$. The SMC has a fixed gain of $\eta = 0.8$.

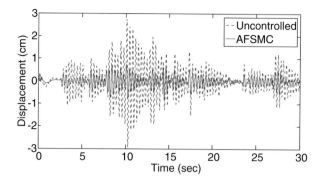

Fig. 6.9 Uncontrolled and controlled displacements of the top floor using AFSMC

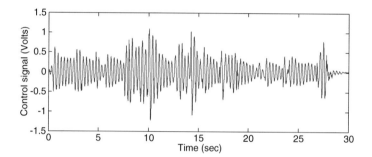

Fig. 6.10 Control signal from PID controller

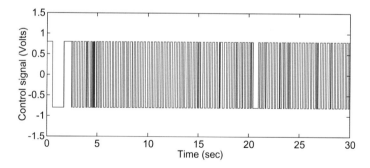

Fig. 6.11 Control signal from SMC

The AFSMC parameters are $\bar{\eta} = 50$, $\eta_{t=0} = 0.8$, and $\mu = 0.001$. These parameters are selected in such a way that a satisfactory chattering and vibration attenuation is achieved. In the case of structural vibration control, the parameter δ indicates the maximum acceptable vibration, which is $0.15\,\text{m/s}^2$. In this experiment, the velocity and position are kept within an acceptable zone by choosing $\delta = 0.02$.

The control performance is evaluated in terms of their ability to reduce the relative displacement of each floor. The wind force signal shown in Fig. 6.6 is used as the excitation signal for the building prototype. Figures 6.7, 6.8, and 6.9 show the time responses of the sixth floor displacement for both the controlled and uncontrolled

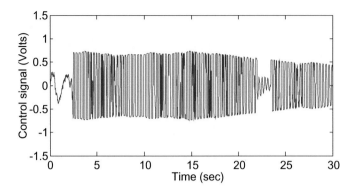

Fig. 6.12 Control signal from AFSMC

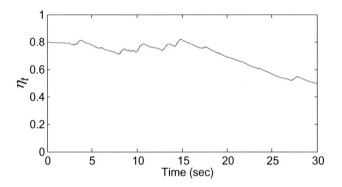

Fig. 6.13 Switching gain adaptation

Fig. 6.14 Uncontrolled and
controlled displacements of
the top floor using PID
controller

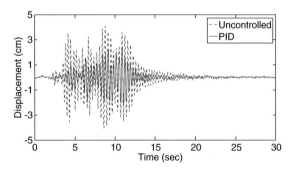

cases. The control algorithm outputs are shown in Figs. 6.10, 6.11, and 6.12. The
gain adaptation of the AFSMC is shown in Fig. 6.13.

From Fig. 6.10, it can be noted that the PID controller generates peak control
signals, and moreover its response time is slower than that of the aggressive SMC.
For normal SMC, increasing η beyond 0.8 resulted in unwanted vibration in certain
points, which is caused by the chattering effect. In order to avoid this problem, the
switching gain is set to $\eta = 0.8$. Among these three controllers, AFSMC achieves the

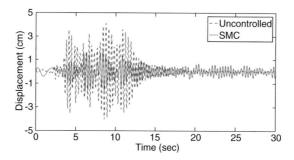

Fig. 6.15 Uncontrolled and controlled displacements of the top floor using SMC with $\eta = 1$

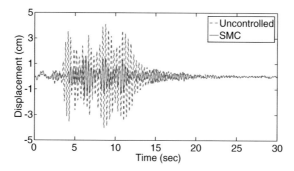

Fig. 6.16 Uncontrolled and controlled displacements of the top floor using SMC with $\eta = 0.8$

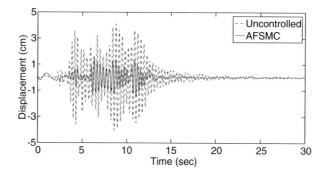

Fig. 6.17 Uncontrolled and controlled displacements of the top floor using AFSMC

better vibration attenuation. The adaptive algorithm of AFSMC significantly reduces the switching gain when the vibration is within the acceptable range, see Fig. 6.13. As a result, both the movement and power requirements of the AMD have been minimized, when compared to the nonadaptive case, see Figs. 6.11 and 6.12.

It is worth verifying the control performance of the AFSMC for an earthquake excitation. Figures 6.14, 6.15, 6.16, and 6.17 show the time responses of the sixth floor displacement for the earthquake excitation. The control algorithm outputs are shown in Figs. 6.18, 6.19, 6.20, and 6.21. A saturation of 1.5 V has been added

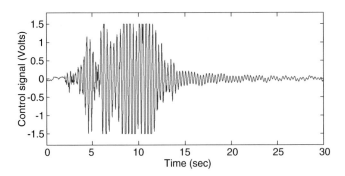

Fig. 6.18 Control signal from PID controller

Fig. 6.19 Control signal
from SMC with $\eta = 1$

Fig. 6.20 Control signal
from SMC with $\eta = 0.8$

Fig. 6.21 Control signal
from AFSMC

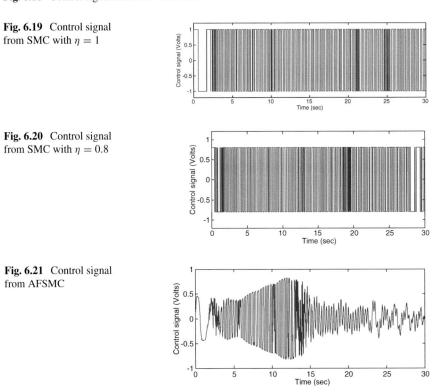

to the PID controller output to avoid any excessive damper movement. Since the earthquake excitation frequency is higher than that of the wind, the AFSMC gain must adapt quickly. The gain adaptation speed is increased by selecting $\bar{\eta} = 65$. The gain adaptation of the AFSMC for the earthquake vibration attenuation is shown in Fig. 6.22.

Figures 6.14, 6.15, 6.16, and 6.17 evidently indicates that the displacement has been reduced considerably using the AFSMC. For normal SMC, increasing η beyond

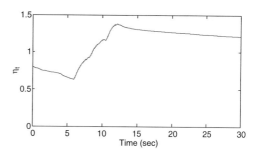

Fig. 6.22 Switching gain adaptation

0.8 resulted in unwanted vibration in certain points, which is caused by the chattering effect. For example, the classic SMC with $\eta = 1$ performs poor while x starts damping from a large to a small value. In Fig. 6.15, after 18 s we can notice an increase in the vibration level. This is due to the fact that SMC switches aggressively with a gain of η (Fig. 6.19), even though the actual vibration is considerably small. This will cause the actuator to add excessive force on the structure. In order to avoid this problem, the switching gain is set to $\eta = 0.8$. Whereas Fig. 6.17 proves that the gain of the AFSMC adapts in a suitable way, the control action is improved near the sliding surface.

References

1. G.W. Housner, L.A. Bergman, T.K. Caughey, A.G. Chassiakos, R.O. Claus, S.F. Masri, R.E. Skelton, T.T. Soong, B.F. Spencer, J.T.P. Yao, Structural control: past, present and future. J. Eng. Mech. **123**, 897–974 (1997)
2. S. Thenozhi, W. Yu, Advances in modeling and vibration control of building structures. Ann. Rev. Control **37**(2), 346–364 (2013)
3. R. Adhikari, H. Yamaguchi, T. Yamazaki, Modal space sliding-mode control of structures. Earthq. Eng. Struct. Dyn. **27**, 1303–1314 (1998)
4. M. Allen, F.B. Zazzera, R. Scattolini, Sliding mode control of a large flexible space structure. Control Eng. Pract. **8**, 861–871 (2000)
5. S.M. Nezhad, F.R. Rofooei, Decentralized sliding mode control of multistory buildings. Struct. Des. Tall Spec. Build. **16**, 181–204 (2007)
6. V.I. Utkin, *Adaptive Sliding Mode Control*, Advances in Sliding Mode Control (Springer, Berlin Heidelberg, 2013)
7. X. Yu, Sliding-mode control with soft computing: a survey. IEEE Trans. Ind. Electron. **56**, 3275–3285 (2009)
8. O. Yakut, H. Alli, Neural based sliding-mode control with moving sliding surface for the seismic isolation of structures. J. Vibr. Control **17**, 2103–2116 (2011)
9. Z. Li, Z. Deng, Z. Gu, New Sliding Mode Control of Building Structure Using RBF Neural Networks, in *Chinese Control and Decision Conference* (2010), pp. 2820–2825
10. H. Alli, O. Yakut, Fuzzy sliding-mode control of structures. Eng. Struct. **27**, 277–284 (2005)
11. S.B. Kim, C.B. Yun, Sliding mode fuzzy control: theory and verification on a benchmark structure. Earthq. Eng. Struct. Dyn. **29**, 1587–1608 (2000)
12. A.P. Wang, Y.H. Lin, Vibration control of a tall building subjected to earthquake excitation. J. Sound Vib. **299**, 757–773 (2007)

13. A.P. Wang, C.D. Lee, Fuzzy sliding mode control for a building structure based on genetic algorithms. Earthq. Eng. Struct. Dyn. **31**, 881–895 (2002)
14. R. Adhikari, H. Yamaguchi, Sliding mode control of buildings with ATMD. Earthq. Eng. Struct. Dyn. **26**, 409–422 (1997)
15. R. Guclu, Sliding mode and PID control of a structural system against earthquake. Math. Comput. Model. **44**, 210–217 (2006)
16. J.N. Yang, J.C. Wu, A.K. Agrawal, S.Y. Hsu, Sliding mode control with compensator for wind and seismic response control. Earthq. Eng. Struct. Dyn. **26**, 1137–1156 (1997)
17. B.F. Spencer, M.K. Sain, Controlling buildings: a new frontier in feedback. IEEE Control Syst. Mag. Emerg. Technol. **17**, 19–35 (1997)
18. Y.K. Wen, Method for random vibration of hysteretic systems. J. Eng. Mech. **102**, 249–263 (1976)
19. F. Ikhouane, V. Mañosa, J. Rodellar, Dynamic properties of the hysteretic Bouc-Wen model. Syst. Control Lett. **56**, 197–205 (2007)
20. J. Resendiz, W. Yu, L. Fridman, Two-stage neural observer for mechanical systems. IEEE Trans. Circuits Syst. Part II **55**, 1076–1080 (2008)
21. J.M. Angeles, L. Alvarez, 3D identification of buildings seismically excited, in *Proceedings of the 2005 IFAC World Congress* (Prage, Czech Republic, 2005)
22. A. Levant, Sliding order and sliding accuracy in sliding mode control. Int. J. Control **58**, 1247–1263 (1993)
23. F. Plestana, Y. Shtessel, V. Bregeaulta, A. Poznyak, New methodologies for adaptive sliding mode control. Int. J. Control **83**, 1907–1919 (2010)

Chapter 7
Conclusions

There has been a large amount of increased research in structural vibration control in the past few decades. A number of control algorithms and devices have been applied to the structural control applications. Linear controllers were found to be simple and effective. More advanced controllers have improved the performance and robustness. Even though this field is well developed, there is still room for further research.

In this book, an active vibration control system for building structures was developed. The system uses accelerometers for measuring the building floor acceleration. However, the accelerometer output signal is polluted with DC offset and other low-frequency noise signals. Direct integration of this signal will result in an inaccurate velocity and position estimation. A numerical integrator was developed, which has different filtering stages to attenuate the noise present in the measured acceleration signal. Experiments showed that the proposed integrator estimates the position and velocity with a good accuracy.

Three different control algorithms are developed for the structure vibration attenuation. Both the classic PID and fuzzy logic control techniques are used. The PID is used to generate the control signal to attenuate the vibration and the fuzzy logic is used to compensate the uncertain nonlinear effects present in the system. The PID gains are selected such that the system is stable in Lyapunov sense. An adaptive technique was developed for tuning the fuzzy weights to minimize the regulation error.

The proposed algorithms are experimentally verified in a lab prototype. The numerical integrator was used to estimate the velocity and position for the controller. Initially, the adaptive Fuzzy PD/PID controllers under seismic excitation were tested. The AFSMC was used to attenuate the wind-induced vibrations in tall buildings. An AMDhas been used to generate the force required to nullify the effects caused by the

© The Author(s) 2016
W. Yu and S. Thenozhi, *Active Structural Control with Stable
Fuzzy PID Techniques*, SpringerBriefs in Applied Sciences and Technology,
DOI 10.1007/978-3-319-28025-7_7

external excitations. The results of the experiments show that both the controllers can attenuate the vibrations considerably well. Also the controllers, especially the AFSMC, can function with nonlinear and uncertain systems like the real building structures.